T0275594

Global Engineering Ethics

Global Engineering Ethics

Heinz C. Luegenbiehl

Rockwell F. Clancy

Butterworth-Heinemann
An imprint of Elsevier

Butterworth-Heinemann is an imprint of Elsevier
The Boulevard, Langford Lane, Kidlington, Oxford OX5 1GB, United Kingdom
50 Hampshire Street, 5th Floor, Cambridge, MA 02139, United States

Notices
Knowledge and best practice in this field are constantly changing. As new research and experience
broaden our understanding, changes in research methods, professional practices, or medical
treatment may become necessary.

Practitioners and researchers must always rely on their own experience and knowledge in
evaluating and using any information, methods, compounds, or experiments described herein. In
using such information or methods they should be mindful of their own safety and the safety of
others, including parties for whom they have a professional responsibility.

To the fullest extent of the law, neither the Publisher nor the authors, contributors, or editors,
assume any liability for any injury and/or damage to persons or property as a matter of products
liability, negligence or otherwise, or from any use or operation of any methods, products,
instructions, or ideas contained in the material herein.

Library of Congress Cataloging-in-Publication Data
A catalog record for this book is available from the Library of Congress

British Library Cataloguing-in-Publication Data
A catalogue record for this book is available from the British Library

ISBN: 978-0-12-811218-2

For information on all Butterworth-Heinemann publications
visit our website at https://www.elsevier.com/books-and-journals

Working together
to grow libraries in
developing countries

www.elsevier.com • www.bookaid.org

Publisher: Matthew Deans
Acquisition Editor: Brian Guerin
Editorial Project Manager: Edward Payne
Production Project Manager: Kiruthika Govindaraju
Cover Designer: Greg Harris

Typeset by SPi Global, India

Contents

3. Engineering Professionalism and Professional Organizations

4. Basic Ethical Principles for Global Engineering

7 Cross-Cultural Issues: Their Importance to Global Engineering Ethics

8 Autonomy

11   The Rights of Engineers

Acknowledgments

The following text is a collective work in the best sense of the term, a collaboration of individuals at different types of institutions worldwide. For that reason, many contributions should be acknowledged.

Texts included in this book were developed, used, and refined in courses at the Rose-Hulman Institute of Technology, Terre Haute, IN, the United States; Kanazawa Institute of Technology, Nonoichi, Ishikawa, Japan; National Yunlin University of Science and Technology, Douliu, Yunlin, Taiwan; the University of Michigan-Shanghai Jiao Tong University Joint Institute, Shanghai, China; and Purdue University, West Lafayette, IN, United States.

At those institutions, the authors would like to thank all of their colleagues, especially Samuel Hulbert, Thomas Mason, James Eifert, Peter Parshall, Toshiyuki Yamamoto, Scott Clark, Pei-Feng Chang, Peisen Huang, Xinwan Li, Gang Zheng, David Barnet, David Hung, Joel Ebarb, Christopher Yeomans, Daniel Smith, Daniel Kelly, Carla Zoltowski, Andrew Brightman, Brent Jesiek, and Qin Zhu.

Many of the case studies began as projects by outstanding students in "Professional Ethics" and "Orienting Engineering Ethics in Terms of China and Chinese Values" at the University of Michigan-Shanghai Jiao Tong University Joint Institute and "Global Moral Issues for Engineers" at Purdue University. These contributions are as follows: Fan Yiming, Shen Li, Xie Yuan, and Yang Guang worked on "The Überlingen Midair Collision"; Kendall Kyle worked on "Engineering a Better Global Future" and "Aligning International Responsibility with Local Needs"; Jiang Weifu, Song Shiwei, and Wei Yi worked on "McDonnell and Miller, and the ASME"; Fan Yilang and Anant Dugar worked on "Global Professionalism?"; Gong Yilan, Zhang Guangyu, Ding Shuhui, Wang, and Zhu Sheng worked on "Development and its Broader Contexts"; Shen Haishangyang, Wang Mengchen, Yan Yifei, Yuan Peng, Chen Yinfan, Dong Yi, Tian Runfeng, and Anant Dugar worked on "Volkswagen, the Environment, and Global Impact"; Zhai Zimu, Song Han, and Chen Mo worked on "The Uber Rape Scandal"; and Qian Shiyi, Tao Zui, and Wei Shupeng worked on "Engineering and Public Knowledge."

James Mollison helped to collect references, offering suggestions for the improvement of the text, and Jessica Sun read through the entire text, offering suggestions for its improvement as well. Additionally, Stephanie Cernera, Mukunda Aithal, Ren Xiang, and Parker Lewellen provided excellent feedback on various parts and suggestions for their improvements.

Finally, Heinz Luegenbiehl would like to thank Jun Fudano, Tokyo Institute of Technology and formerly of Kanazawa Institute of Technology, for the opportunity to pioneer teaching engineering ethics in Japan and for numerous occasions of financial support to conduct research. Rockwell Clancy would like to thank, first and foremost, Heinz Luegenbiehl, from whom he has learned a tremendous amount about engineering ethics, especially in international contexts. Additionally, he would like to thank the University of Michigan-Shanghai Jiao Tong University Joint Institute for a teaching leave and Purdue University for a grant during the spring 2016 semester to work on this book and, last but not least, his family, friends—especially Kelly VandenBosche and Arjen Kleinherenbrink—and students for all their support over the years.

Chapter 1

Introduction: Engineering Ethics from a Global Perspective

Chapter Objectives

Having read this chapter and answered the associated questions, readers should be able to

- describe recent changes that have taken place in the field of engineering, and why this necessitates approaching engineering ethics from a different perspective;
- articulate not only the nature of ethics in general but also why it should be of particular concern to engineers;
- explain the problem of theory and interconnected roles that reason, engineers' role responsibilities, and case studies play in approaching engineering ethics from a global perspective.

CASE STUDY ONE—THE ÜBERLINGEN MIDAIR COLLISION: SYSTEMS CONFLICTS AND GLOBAL CONTEXTS

At approximately 9:35 p.m. on Jul. 1, 2002, two planes collided midair near Überlingen, Germany, killing all 71 persons on both planes. A number of human-controlled and human-automated technological systems were in place to avoid such an incident ever occurring. This collision was, thus, the result of a confluence of circumstances and conditions that illustrate the increasingly complex and global contexts of technology in modern life. Examining this case helps to introduce readers to this context and its associated problems.

The Überlingen collision occurred between a Tupolev 154 passenger plane traveling from Moscow, Russia, to Barcelona, Spain (henceforth 154), and a Boeing 757 cargo plane traveling from Bergamo, Italy, to Brussels, Belgium (henceforth 757). The pilots of both planes were well-trained, seasoned flyers. The pilots of 154 to Barcelona were Russian, and the pilots of 757 to Brussels were British and Canadian. Their nationalities are important, as we will see shortly, since training and cultural differences contributed to the collision.

Global Engineering Ethics. http://dx.doi.org/10.1016/B978-0-12-811218-2.00001-1
1

757 and 154 began communicating with Swiss air traffic controller Peter Nielsen, of the Zurich area control center at 9:21 p.m. and 9:30 p.m., respectively. Nielsen told 757 to climb to flight level 360, which it did, and, shortly after, 154 entered Zurich airspace at that same flight level. The pilots of 757 failed to report to Nielsen that they had ascended to flight level 360. When 154 entered Zurich airspace, Nielsen told the pilots to change their radio frequency to avoid interference, since 154 was initially using the same radio frequency as 757. Nielsen was unaware that the two planes were at the same flight level, and his attention was diverted for two main reasons.

First, although two air traffic controllers were working that night, one was on break at the time, leaving Nielsen by himself. Nielsen was thus in charge of two navigation stations—one for high altitudes on which 154 and 757 appeared and one for Friedrichshafen and St. Gallen-Altenrhein airports—with their own radios and screens; the screens were located a meter apart from one another, forcing Nielsen's attention to be divided between them. After instructing the pilots of 154 to change their radio frequency, Nielsen turned his attention to the Friedrichshafen and St. Gallen-Altenrhein airports screen, directing an Airbus approaching German airspace.

Second, the main telephone system used to communicate with nearby airports in Germany was—unbeknownst to Nielsen—down for maintenance, and the backup telephone system was not working as a result of software failure. Whereas Nielsen was not informed about the first, no one knew about the second. Again, after instructing the pilots of 154 to change their radio frequency, Nielsen was preoccupied with the phone system, attempting to communicate with air traffic controllers in Germany regarding the approaching Airbus. Additionally, although German air traffic controllers recognized the potential collision between 757 and 154, since both the main and backup phone systems were down, they were unable to communicate this information to Nielsen. Despite the failure of 757 to report its ascent, inattention and preoccupation by Nielsen, and failure of the main and backup phone systems, air travel and traffic control systems have built-in redundancies.

Despite the failures mentioned so far, there were two automated technologies in place that should have prevented the Überlingen collision: the short-term conflict alert (henceforth STCA) and the traffic collision and avoidance system (henceforth TCAS). The STCA is an automated alarm system that alerts controllers 2 min before any potential collision. Hence, 2 min before the Überlingen collision, this warning system should have alerted Nielsen that 757 and 154 were dangerously close. On the night of the collision, however, this system was down.

The TCAS is installed on planes, and it alerts pilots and provides instructions on how to avoid potential collisions. At 9:34:42 p.m., the TCAS alerted the pilots of 154 and 757 of the potential collision and then, at 9:34:56 p.m., directed the planes to ascend and descend, respectively. At this point, if the pilots of both planes had followed these instructions, then the tragedy could have been avoided. However, although the TCAS alerted the pilots of the danger and issued instructions on how to avoid a collision, this information was not relayed to the air traffic controller, Nielsen.

Hence, 7 s after the TCAS initially alerted the pilots of 154 and 757, Nielsen recognized the danger and directed 154 to descend to flight level 350. His instructions were the opposite of those the pilots of 154 received from the TCAS, a conflict between human instructions and an automated system. However, the Russian training of the 154 pilots had instructed them to follow the directions of air traffic controllers, and 154 descended accordingly. There were no mechanisms or safeguards in place if the pilots failed to comply with the TCAS instructions. It should also be noted that just over a minute (1:10) passed between the time the TCAS alerted the pilots to the potential collision and the time the collision occurred. Little time was thus available for either the pilots or Nielsen to gain their bearings.

The TCAS on 757 instructed the pilots to descend more, which they did and reported to Nielsen, although they received no response. With the TCAS on 154 continuing to instruct the pilots to climb, it wasn't until 9:35:27 p.m. that 154 stopped descending and started climbing. By this time, however, it was too late, and the two aircrafts collided at 9:35:32 p.m.[1]

This case raises a number of interesting questions regarding the complex, global contexts of technology in modern life: which persons and systems contributed to the Überlingen collision? Where does primary responsibility lie? If not one person, then multiple persons? Are persons primarily to blame for this collision? If not, then where should blame be placed? What could have been done differently to avoid this collision, and what can be done to prevent such tragedies from occurring in the future? Who's responsible for the ways modern technology is used? What role did the interaction between human beings and technology play in this collision, and what does this tell us about technical and organizational designs and their implementations? These are difficult questions, the answers to which are by no means entirely clear. For precisely those reasons, these are some of the issues considered going forward.

1.1 WHAT'S CHANGED AND WHY IT MATTERS: INITIAL ASSUMPTIONS[2]

Traditional engineering practice has been relatively localized to specific cultural contexts. Therefore, ethics education for engineering students could legitimately be based on the background conditions existing in a particular society, and instructors could assume general familiarity with these among students.

1. This account is based on Bundesstelle für Flugunfalluntersuchungg (2004), Johnson (2004), Nunes and Laursen (2004), and Turney (2007).
2. Materials in this chapter previously appeared in Luegenbiehl (2010).

In the latter part of the 20th century and into the 21st century, however, this was no longer the case.

The previously existing conditions underwent significant changes, including the coming to dominance of multinational corporations, the location of plants by national corporations in other countries, the increasing international mobility of engineers, and the establishment of international supplier and customer systems. While some texts on engineering ethics ignored these developments, others have responded by adding one or more chapters regarding issues sometimes encountered by engineers dealing with foreign entities, such as questions of different ethical and religious systems and grease payments.[3] Such responses, while legitimate first attempts to deal with the new international environment of engineering, do not address the need for a fundamental reconceptualization of how ethics should be conceived and taught, assuming it is no longer sufficient to look at international issues from within the framework of national perspectives.

This text seeks to rethink engineering ethics at a more fundamental level, using the global environment of engineering as the starting point rather than a mere addition. Toward this end, it is necessary that the particular national assumptions about the practice of engineering and theoretical foundations of ethics arising from specific cultural traditions be set aside. Only those assumptions that can be justified based on the nature of engineering itself and universal human characteristics should be used as a starting point. This is important as those assumptions would be ones on which all could agree. The following are the most significant of these:

1. The first assumption concerns the nature of engineering and the world: Engineering is not value neutral, and the activities of engineers should not leave the world less well-off than it was before. A cursory definition of "engineering" could be the following: *the transformation of the natural world, using scientific principles and mathematics, in order to achieve some desired practical end.* Chapter 4 returns to different understandings of engineering and further justifies the definition given here. This initial assumption is important when considering engineering in the context of business environments, since no one wants to buy or pay for things and services that ultimately hurt him or her. The question then becomes what limits should be placed on engineering processes, and what justifies these limits. This amounts to requiring that the benefits of engineering to the world outweigh the costs. Further, no cost should be catastrophic in nature, as that would make it difficult to measure against potential benefits.

2. The second assumption concerns the nature and use of reason: the ability to use reason is a relatively universal human characteristic; its main properties

---

3. Although the use of footnotes and references has been kept to a minimum, the authors want to direct interested readers to resources in the fields of engineering ethics, ethics, philosophy, science, technology, and society studies, and psychology that would be relevant. A few of the major texts in engineering ethics to date include Harris, Pritchard, Rabins, James, and Englehardt (2013), Whitbeck (2011), Van de Poel and Royakkers (2011), and Martin and Schinzinger (2010).

are shared in common among all human beings. Reason is generally defined as the process of logical, discursive thought, being able to provide clear justifications for decisions. Although not all human beings possess reason or evidence reasonableness all the time, engineers—as a result of their chosen profession—should be especially committed to the use of reason.

3. The third assumption concerns human nature and economics: human beings exhibit a tendency to seek out their own gain, although not exclusively so, and the dominant manifestation of this tendency in the contemporary world has come about through the adoption of some form of capitalistically motivated action. In part, this explains why the activity of business will be stressed in a text on engineering ethics, discussed at length in Chapter 4. This assumption does not mean, however, that human beings cannot or will not seek out the benefits of others, acting in an altruistic fashion. Chapter 3 examines the nature of professions and professionalism. There it is explained that, as professionals, engineers should actively promote the well-being of not only the professional community but also that of society as a whole.

4. The fourth assumption concerns cultures and values: different cultures have, at a fundamental level, different value systems, and these may conflict with those of other cultures. These systems can change, and no one is justified in assuming any one of these value hierarchies is necessarily the correct one or better than others. Chapter 7 explains and examines these assumptions regarding the natures of and relations between cultures and values in greater depth.

5. The fifth and final assumption concerns the nature of ethics and religion: a secular approach is the most appropriate for the study of global engineering ethics. In many societies, there is a close connection between ethics and religion, to the extent that, in some, religious perspectives dominate all discussions about morality and its justification. The problem with this is that there are a number of major religions whose ethical commitments concerning how people should act tend to be closely tied to their metaphysical commitments regarding the nature of reality. This means that what, at one level, could be very similar ethical claims are, at another level, interpreted differently, precisely because of their differing religious contexts. Thus, religion exhibits a tendency toward obstructing possible agreement regarding ethics among different peoples of the world. As a global standard of ethical conduct for engineers is sought here, the religious perspective will be set aside, while at the same time recognizing its importance to the daily lives of people in different cultures. This assumption touches on the difference between one's personal ethical commitments and those that follow from one's professional role:

- How do you think your religion—or lack thereof—influences your ethical outlook?
- Do you agree with the assumptions outlined above? Why or why not? If you disagree, then what would you change?

1.2   WHAT IS ETHICS?

Here the terms "ethics" and "morality" will be used interchangeably. A variety of different ways of defining these terms are in common use, but many of them rely on technical differences in interpretations. The definition proposed for use in this text is that *ethics is about actions that have the potential to seriously impact the lives of others.* The meaning of the word "serious" is, of course, vague but will be clarified in subject matter analysis.

It should be noted that, for the purposes of this text, ethics concerns human beings, although significant discussion exists about whether or not that should be its limit. For example, some people have argued that animals should have moral rights, based on their capacity to suffer. Our limitation does not mean, of course, that either animals or the environment should not be discussed ethically, since our actions toward other beings and things often have consequences for the lives of human beings. Here global warming as such, for instance, would not be of ethical interest, but considering that what people do to the environment can seriously harm human beings, it becomes a matter of ethical concern. Many discussions in ethics also take place regarding who is to be defined as a "person," but the theoretical ramifications of that issue will not be explored here.[4]

It should also be noted that, according to this definition, ethics is concerned with potential effects on others. Some ethical theories have proposed that ethics is about the furtherance of self-interest. As people do not require encouragement to seek out and further their own self-interests, the position here is that there is no need to develop rules about behaviors of this sort—human beings do so quite naturally.[5] Much of ethics is ultimately about setting restrictions on or limits to behaviors.

Especially in the context of a global ethics, it must also be noted that this definition of ethics is framed in terms of actions. Questions regarding the character of individuals, for example, are thus left unresolved. There exist a number of ethical traditions that stress character development or spiritual state of persons. Neglecting this issue in framing an ethics for engineers is not to indicate that this is an unimportant dimension of ethics but, rather, that it is often not suitably assessed by an outsider.[6]

---

4. For discussions of such issues, see Smith (2010). Regarding animal rights and ethics, see Singer (1975), Rowlands (2002), and Carruthers (1992). Concerning environmental ethics, see Naess (1973), O'Neill (1992), and Jamieson (2001). This issue is taken up further in Chapter 11, regarding the nature of rights.

5. The position that ethics *is* about the furtherance of self-interest is known as "descriptive" egoism. The position that ethics *should* be about the furtherance of self-interest is known as "normative" egoism. For more on descriptive egoism, see Feinberg (1999), Mercer (2001), and Rachels (2003). For an overview of normative egoism, see Shaver (2015).

6. Stemming from the thought of ancient Greek philosopher Aristotle, ethical questions regarding the character of individuals are included under the broad purview of "virtue" ethics. Concerning recent psychological and philosophical criticisms of the understanding of moral psychology on which this ethical position is based, see Wilson (2004) and Doris (2005).

A further concern is with the distinction between what has been called the descriptive and normative dimensions of ethics. "Descriptive" ethics refers to the way individuals *actually* act or rules that exist in a society. "Normative" ethics refers to how people *should* act, independently of how people actually act and existing social rules. Because this text is concerned, in part, with taking into account particular local conditions in establishing norms for engineers, it proposes an interplay between the descriptive and normative dimensions of ethics. This would include the frank recognition of cultural and regional value hierarchies—for example, the awareness that some cultures place greater emphasis on loyalty than honesty, and others place greater emphasis on honesty than loyalty—in identifying and resolving ethical issues. Chapter 2 outlines this process in detail. Thus, unlike some philosophical approaches to ethics, it holds that engineers must account for real-world conditions.[7]

Given the emphasis on cultural diversity, it is especially important to recognize the limited scope of ethics. In the first place, actions that affect only one's self will not be considered in the scope of ethics. For example, the private thoughts of an individual—as repulsive as those thoughts might be if they were known by others—if they do not have effects on others, will not be considered within the scope of ethics. In the second place, matters of courtesy or custom—although others not following them may cause offense—will not be considered within the scope of ethics. For example, although I might recoil in horror if you fail to take off your shoes when entering my house, I would not be seriously harmed by this action:

- Do you think animals should be considered in ethical discussions? Why or why not?
- Do you think human beings should consider their effects on the environment, beyond merely how these effects affect human life? Explain your answer.

1.3 WHY ETHICS FOR ENGINEERS?

Matters of ethics are—or at least should be—of concern to everyone. In order for people to exist within communities, including the global community, there must be rules to prevent destructive behaviors toward others. Otherwise, the world would descend into chaos. Laws help to establish such rules, and they serve as a foundation for the enforcement of these rules.

However, as will be explained more fully in Chapter 10, laws are insufficient for this purpose—at least in part because laws are politically arrived at instruments. Due to this fact, laws might even contradict generally accepted ethical

---

7. This approach is thus informed by advances in "behavioral ethics" and "experimental philosophy." For readable accounts of recent advances in these fields, see Bazerman and Tenbrunsel (2013) and Alexander (2012), respectively.

norms. Ethics is thus necessary both as a supplement and counterbalance to law. For example, to make it illegal to ever lie to others would be an unenforceable law, and, thus, laws typically restrict themselves to making it illegal to lie under oath or to an official authority. Some traditional ethical theories would claim lying is wrong even in conditions that cannot be covered by law, thereby covering a greater range of actions than is possible by law. Further, illegality is not the same as unethicality. For example, although it is illegal to drive one's car on the left-hand side of the road in the United States—and similarly illegal to drive one's car on the right-hand side of the road in the United Kingdom—before the law has been established and promulgated, there is nothing inherently more ethical about driving one's car on the right- or left-hand side of the road. Finally, as political instruments, laws sometimes reflect the interests of the powerful, as opposed to the interests of justice. Thus, laws can contradict the requirement not to seriously harm others, for instance, through the establishment of institutions such as slavery and serfdom.

If ethics is or should be of concern to everybody, however, the question still remains of why ethics is and should be of special concern to engineers: why not simply offer a course in ethics rather than engineering ethics? There are several reasons.

The first reason is that engineers are singularly important to the world. Engineers themselves are often unaware of the powers they exercise, of their abilities to affect the lives of others. Instead, they tend to think of themselves as small cogs in big machines, with little influence over the course of events. A major purpose of engineering ethics could, thus, be termed "consciousness raising," helping engineers to become aware of their powers and the responsibilities that go along with the exercise of these powers.[8]

The second reason is that engineers are, in fact, different from the general public. They have specialized knowledge and skills that are only acquired and acquirable through long periods of intensive study and training. These characteristics are associated with professions and professionalism, about which more is said in Chapter 3. To the public, the knowledge of engineers is the equivalent of a black box: what goes on inside that box is a mystery. The public thus exists in a relationship of dependence on engineers. It must trust the work engineers do on its behalf. Through the study of ethics, engineers can help assure the public that they are worthy of this trust.

The third reason is that several specific ethical obligations for engineers are derived from the nature of engineering itself. That means that these obligations are ethical duties specific for engineers that are not applicable to all persons, only those in the specific roles of engineers. A further explanation of this basis, these principles, and their justification comprises the largest part of this text, in Chapters 3–10. A discussion of general ethics would, thus, be insufficient for understanding ethical concerns within engineering.

---

8. Regarding the way that unethical actions result more from "blind spots"—biases constituting human psychology that make it difficult for persons to realize they are confronting ethical dilemmas—than reflectively making poor ethical decisions, see, again, Bazerman and Tenbrunsel (2013).

The final reason is that the positions of engineers entail not only responsibilities but also rights, explained in Chapter 11. To exercise these appropriately, on behalf of themselves and the public, engineers must first become aware of these responsibilities and rights. Unfortunately, due to their work in business environments, engineers might sometimes be subject to undue and unjustified pressures, to which they might respond inadequately without the ability to justify their decisions:

- What advantages would an engineer trained in global ethics have over an engineer who has no such ethical training? Explain your answer.

1.4 A GLOBAL PERSPECTIVE

Given the role of engineers and engineering in the contemporary world, here a global approach to engineering ethics is taken. While this might make sense to students growing up now, it is, in fact, a departure from the tradition.

Engineering ethics is an area of study that originated in the United States. Because the US model of engineering ethics was first, it has become somewhat of a global standard.[9] The problem is that this model carries within it some uniquely US or Western features, such as an emphasis on ethical theory and the ideal of professions. Some of these features are neither readily nor appropriately adaptable to other parts of the world.[10]

It is thus necessary to rethink engineering ethics from the ground up, which is the approach taken here. One could, of course, begin with a cultural perspective other than that of the United States, but that would raise the same issues of being too focused on one cultural perspective at the expense of others.[11] Consequently, the approach here consists in beginning to study engineering ethics without any cultural presuppositions, bringing in questions of culture later. This is the meaning of a "global perspective" on engineering ethics.

As will become clear in the remainder of the text, a number of important consequences follow from the adoption of a global approach. In the first place, no current perspective on engineering ethics is privileged simply because of its origin. Rather, of importance is that discussions follow from the assumptions laid out. Further, there will be a lack of emphasis on traditional Western ethical theories, since these are not shared in common on a global basis. In thinking about engineering ethics independently of a particular society, it might instead be helpful to think about engineers themselves as a community with a shared set of values. A bond is created among engineers based on these values, just as there typically exists a set of common core values in other societies. These values

9. Concerning the history of engineering ethics, see Davis (1995).
10. With regard to the nature of professionalism specifically, see Luegenbiehl and Fudano (2005) and Didier (2010).
11. Regarding the contributions of Chinese and—more broadly—"Eastern" values to engineering and engineering ethics, see, for instance, Chang and Wang (2008) and Wang and Zhu (2012).

provide for an element of trust and legitimate defense against forces that would undermine the appropriate application of technology. A major task here will be to enunciate the values suitable for global engineers.

1.5 FOUNDATIONS FOR THE ANALYSIS OF ETHICAL ISSUES

1.5.1 Problems of Theory: Theoretical and Cross-Cultural Disagreements

Many engineering ethics texts currently available begin with a discussion of ethical theories. Such theories have been developed over several centuries, to serve as foundational principles for ethical analysis. Dealing with specific ethical issues that arise in the world has, consequently, become known as "applied ethics," indicating that, procedurally, it is the ethical principles that form the basis of discussion. A move away from this approach has begun, because of difficulties students find in applying these general principles, and because there continues to be serious disagreement among philosophers as to which principle or set of principles should be used.[12]

When ethics is viewed on a global basis, these problems become magnified, since foundational ethical principles as the major source for decision-making are specific to the Western philosophical tradition.[13] Consequently, many students not familiar with that tradition have a difficult time understanding that approach in the first place, before even questioning whether they agree or disagree with a particular instantiation of that approach. In such cultures, courses in applied ethics face the danger of becoming introductions to Western philosophical thought, at the expense of actually considering specific, applied issues. Fundamental disagreements about the appropriateness of utilizing ethical theory—perhaps based on the claim that ethical theory seeks to operationalize a uniquely Western perspective and impose it on the rest of the world—also stand in the way of reaching agreement on how specific ethical issues should be resolved.

In addressing a global audience, therefore, references to specific ethical theories and the identification of duties based on particular religious and philosophical traditions are avoided. Instead, the claims included here are based on three features: human reason, role responsibilities, and case study. There is a universal dimension to be found in this approach. However, it will also be clear to the discerning reader that there are, at times, implicit connections to

---

12. Concerning the move away from professional as "applied" ethics, see Chadwick and Schroeder (2002a).
13. For a very readable but highly informative account of the relation between ethics, advances in cognitive science, and East Asian philosophical and religious thought, see Slingerland (2015), and with regard to the foreignness of the formulation and application of ethical theories to concrete situations within Confucian thought, for example, see Ames (2011) and Hall and Ames (1987).

traditional claims of philosophy and religion.[14] It is left to students more knowledgeable about these subjects to make such connections explicit for themselves, analyzing the adequacy of their use.[15] However, there is not a sense of exclusivity or "final court of appeal," traditionally associated with the use of Western ethical theories and particular religious doctrines. The hope is that the typical engineering student will be spared the need for a separate study of theories in finding a basis for communicating about ethics with other students—and eventually colleagues—from different cultural backgrounds.

1.5.2   The Role of Reason: Its Universality and in Engineering

In general, the ability to reason is a characteristic common to all adult human beings—the ability to logically think through problems. Central characteristics of reason include consistency, reversibility, deliberation, dispassionateness, and universality. So central to human existence is the ability to reason that, some have claimed, it is the core of human identity—what separates humans from other creatures. Whether this is true may be subject to debate, but without the ability to reason, engineering—at least in its modern manifestation—could not exist. Engineers should, therefore, accept the assumption of the validity of an approach to life using reason. They have committed to the use of reason through their chosen professions as engineers. While other elements of humanness are also significant—such as an emphasis on emotions, for example—only reason is expected to manifest itself in a uniform fashion across cultures. Given the above characteristics, the use of reason as an assumption in a cross-cultural and global foundation for the development of ethical principles in engineering seems justified[16]:

● Are there benefits to having different ethical paradigms around the world? Why or why not?

1.5.3   Role Responsibilities: Special Duties

Here the idea of "role responsibilities" is one of the foundations for analysis. Role responsibilities, as the name indicates, are duties associated with a particular position one occupies in life. All of us have many roles, and different

---

14. This would include, for example, the presumed universality of human reason and Western philosophical understandings of personhood in dualistic terms, based in the thought of Plato and Descartes.
15. The basis of these texts have been used in courses on engineering ethics and science, technology, and society studies in the United States, Japan, Taiwan, and mainland China, and the authors encourage both instructors and students to share their experiences, especially when these would prove useful in developing future versions of this text and its associated study materials.
16. Regarding the claim that some types of reason would not be universal, see, for example, Boas (1940), Geertz (1993), and Whorf (1956).

duties may be associated with each of these roles, arising out of the nature of the role itself rather a more general principle. Parents, for instance, have a special responsibility to nourish and support their children, much beyond the responsibility that strangers or the state might have. Similarly, teachers have special responsibilities to ensure that their students learn, and physicians have special responsibilities to the health of their patients.[17]

In making a list of one's roles, it is quickly discovered that these are numerous. An engineer might also be a parent, golfer, sibling, student, customer, driver, and so on. Once one identifies theses roles, two things should occur: first, deciding what special duties are associated with the roles—or at least to be informed by others what these should be—and, second, prioritizing the relationships among the many roles. Chapter 3 begins the process of determining the responsibilities associated with being an engineer, based on the nature of engineering professionalism itself.

The second process of prioritizing among different roles is too complicated to be carried out in the abstract and should also be flexible over time. The recognition that one has many, varied role responsibilities thus points to the need for students of applied ethics to be able to evaluate specific situations on their own, for no entirely pregiven matrix is ever available. This is not to say, however, that one cannot learn from others, from the experiences of those who have come before. In not only engineering ethics but also business education and ethics, this fact provides a large part of the justification for a case-study approach.

1.5.4 Consideration of Cases

Working with specific sets of circumstances requires the ability to analyze a series of events. This fact points to the importance of case analysis in engineering ethics education. Studying a variety of cases serves several important purposes. Analyzing cases

1. emphasizes the process of active learning, which has been shown to facilitate student understanding;[18]
2. helps students to build up an experience base regarding ethical issues before they actually have to deal with these issues in their professional lives;
3. helps students to develop their ability to analyze and solve problems, in terms of not only ethics but also engineering, enabling them see the connections between ethics and engineering more clearly;[19]

---

17. Concerning the relation between professional role responsibilities and ethics in general, see Frankel (1989). Regarding the nature of professional role responsibilities and ethics within the fields of education and medicine, see Strike and Ternasky (1993) and Chadwick and Schroeder (2002b), respectively.
18. Regarding the way active learning facilitates understanding, see Prince (2004).
19. See Whitbeck (2011) regarding the ways that the activities of engineering design and ethical analysis employ similar modes of reasoning that mutually enforce each other.

4. aids in the ability to generalize by looking at a variety of specific instances and, thereby, helps students to form a more general ethical perspective.

1.6 PREVIEW: WHAT'S TO COME...

The remainder of this text seeks to establish the basic elements of a global approach to engineering ethics. In doing so, two important elements are stressed: first, the connection of engineering to the global business environment and, second, the need to understand that a variety of cultural value systems exist in the world, which must be recognized and accounted for. This approach, therefore, asks more of students than that typical of engineering ethics education, where a common set of shared values between the instructor, students, and other students are assumed. It also requires that students place the activity of engineering in a real-world framework, rather than viewing it as an ideal phenomenon that can be analyzed independently of external considerations. While these tasks are challenging, facing them is the best preparation for students to become ethical engineers in the 21st century.[20]

It should also be noted that, in utilizing this approach, at times the text takes a prescriptive tone, asserting that specific obligations and rights for engineers should be agreed to in global contexts. It does not follow from this, however, that the proposals made in the text are beyond debate. In fact, students are encouraged to question the propositions put forward in the text and arrive at their own conclusions regarding specific ethical issues raised. In doing so, they contribute to the field of engineering ethics.

CASE STUDY TWO—ENGINEERING A BETTER GLOBAL FUTURE: FUSION POWER ACROSS BORDERS

After the 1985 Geneva Convention, Mikhail Gorbachev, secretary general of the Soviet Union, and Ronald Reagan, president of the United States, released a statement that "emphasized the potential importance of the work aimed at utilizing controlled thermonuclear fusion for peaceful purposes" and "advocated the widest practicable development of international cooperation in attaining this source of energy" ("The ITER initiative," 2005). This statement signaled to scientists a focus on thermonuclear fusion and launched an international collaboration of immense scale. In the decades since, thousands of scientists and engineers, from 35 countries, have cooperated on the design, management, and construction of an international thermonuclear experimental reactor (henceforth ITER).

---

20. Again, the importance of this ability would consist—first and foremost—in being able to discern the ethical dimensions of engineering and engineering-related situations since, according to Bazerman and Tenbrunsel (2013), the greatest impediment to acting ethically are innate psychological biases that prevent persons from being able to recognize the ethical dimensions of situations.

As the world faces an ever-increasing population and decreasing resources, using new technologies to supply demand becomes correspondingly important. Central to such endeavors is the development of green technologies that would result in less pollution, such as fusion energy. Additionally, since the behaviors of disparate national populations affect each other as never before, coordinating the efforts of citizens, politicians, and scientists on a global scale is necessary. To envision and encourage such large-scale international technology projects, the ITER project can serve as an example, one of the largest global engineering collaborations in history.

Thermonuclear fusion consists in two nuclei colliding and fusing to form a heavier nucleus, releasing an incredible amount of energy in the process. For fusion to occur, particles must be exposed to extremely high temperatures for sufficient periods of time. A tokamak reactor uses powerful magnetic fields to direct particles through its "doughnut-shaped vacuum chamber," which confines these particles for a long enough time, at a high enough temperature, to cause them to collide and fuse ("What is ITER?"). Unlike energy resulting from either fossil fuels or nuclear fission, the waste products of energy resulting from nuclear fusion could be harmless. Upon its completion, the ITER will be the largest tokamak reactor in the world and the first nuclear fusion device to produce more electricity than it consumes. The goal of the ITER project is twofold: producing 500 MW of fusion power and aiding scientists in researching the possibility of commercially viable fusion-based electricity.

The ITER project was officially started in the early 2000s and has involved collaboration between China, India, Japan, Korea, Russia, the United States, and countries in the European Union. The reactor was scheduled for completion in 2016 with an estimated cost of €5 billion. Due to difficulties involved with organizing thousands of people from so many countries, however, now, the completion date has been pushed back to the mid-2020s, and the budget increased to €15 billion. Michael Classens, head of ITER communication, claims, "the main challenge for the project is not the science and technology itself, but the management as a whole, the way these 35 countries cooperate" (Gupta, 2016).

Robert Iotti, chair of the ITER Council, traces these problems to the organization of the council itself: the responsibilities of countries were based on monetary contributions rather than which could best build the required parts. Additionally, the ITER council has little central power. All major decisions essentially need unanimous approval, which requires more time for discussions. The project relies on different nations completing different parts, so pieces of equipment that arrive later cause delays to the entire project, resulting in the later projected completion date.

Ronald Reagan "saw the potential positive impact of American and Soviet engineers living and working together for some period of time" (Degitz, 2015). He promoted the ITER project not only for scientific reasons but also as a kind of social experiment in international collaboration. At the beginning of the project, considerable time was spent just determining a way for different nations to work together. Michael Roberts, chair of the ITER Council Working Group on Export Control, said that "nearly all the articles in the ITER Agreement deal principally with the practical aspects of collaboration at government level and not with the technical aspects of fusion" (Degitz, 2015). Given all the initial work that went into all the parties finding common ground as a way to understand each other and work together, he believes this framework can be applied to future projects in different fields.

Given the ever-increasing need for cross-cultural/cross-national collaborations, the ITER project can serve as an example, "a model for other large international science collaborations" (Arnoux, 2009).[21] Some of the greatest scientific achievements have been the result of teams with diverse backgrounds. In terms of not only their goals but also their administrations, the nature of such projects draws attention to the need for broad but commonly agreed-upon principles with ethical import. Projects like ITER involve peoples from different cultural and national backgrounds, which have the potential to seriously impact—for better and worse—peoples worldwide. These facts highlight the need to consider the ethics of engineering from a global perspective, moving beyond any one cultural or national paradigm:

- Explain how large-scale, cross-cultural/cross-national technology projects such as ITER can help to engineer better futures.
- How is the ITER project both a good and bad example of international collaboration?

1.7 SUMMARY

As the case of the Überlingen midair collision makes clear, the modern contexts in which technology is employed are becoming ever more complex and global, involving human-operated and automated systems worldwide. As a discipline of study and practice, engineering ethics began in the United States and has been largely based on national, Western assumptions—especially those regarding theory-based approaches to ethics. In the 21st century, such an approach is no longer sufficient. A global approach to engineering ethics should base its foundational assumptions on universally shared characteristics regarding the nature of engineering and being human, while at the same time recognizing the role that cultural value differences play in engineering contexts. Central to the approach taken here is the use of reason, role responsibilities of engineers, and case studies, and the global business environment of engineering is an important point of reference as well. The ITER project helps to illustrate the importance of scientists and engineers, politicians, and citizens working together across national and cultural divides to address needs related to sustainable global development. Doing so involves issues of ethical import, highlighting the need to rethink the ethics of engineering from a global perspective.

---

21. This account is additionally based on Arnoux (2015), Clery (2015), Lucibella (2015), and Olivero (2016).

REVIEW QUESTIONS

1. What five assumptions frame this book's treatment of engineering ethics?
2. What is the definition of ethics for the purposes of this text?
3. Describe the differences between "descriptive" and "normative" ethics.
4. Why are laws insufficient in regulating peoples' behaviors, making ethics necessary?
5. Identify and discuss three reasons why engineering ethics should be distinguished from ethics in general.
6. List two issues or complications associated with applied ethics. Explain your answer.
7. List two reasons case studies are helpful in studying engineering ethics. Explain your answer.

REFERENCES

Alexander, J. (2012). *Experimental philosophy: An introduction.* Cambridge: Polity Press.

Ames, R. (2011). *Confucian role ethics.* Honolulu: University of Hawaii Press.

Arnoux, R. (2009). 2001–2007: The making of ITER. *ITER.* 10 July. https://www.iter.org/newsline/89/862.

Arnoux, R. (2015). Conceived in Geneva, born in Reykjavik, baptized in Vienna. *ITER,* 16 November. https://www.iter.org/newsline/-/2323.

Bazerman, M., & Tenbrunsel, A. (2013). *Blind spots: Why we fail to do what's right and what to do about it.* Princeton, NJ: Princeton University Press.

Boas, F. (1940). *Race, language, and culture.* Chicago: University of Chicago Press.

Bundesstelle für Flugunfalluntersuchung. (2004). *German Federal Bureau of Aircraft Accident Investigation Report.* Brunswick: Bundesstelle für Flugunfalluntersuchung. 2 May.

Carruthers, P. (1992). *The animals issue: Moral theory in practice.* Cambridge: Cambridge University Press.

Chadwick, R., & Schroeder, D. (2002a). *Challenges and problems in applied ethics.* In *Applied ethics: Critical concepts in philosophy.* Vol. 1. New York: Routledge.(pp. 167–243).

Chadwick, R., & Schroeder, D. (2002b). Ethical issues in medicine technology and the life sciences I & II. In *Applied ethics: Critical concepts in philosophy.* Vols. 2 & 3. New York: Routledge.

Chang, P., & Wang, D. (2008). Contribution of Chinese values for the quality standards of global ethics across the nation borders. In *International conference on engineering education. New challenges in engineering education and research in the 21st Century. 27–31 July Pécs-Budapest, Hungary.* http://www.ineer.org.

Clery, D. (2015). ITER fusion project to take at least 6 years longer than planned. *ITER,* 19 November. http://www.sciencemag.org/news/2015/11/iter-fusion-project-take-least-6-years-longer-planned.

Davis, M. (1995). An historical preface to engineering ethics. *Science and Engineering Ethics, 1,* 33–48.

Degitz, L. (2015). Michael Roberts: We had a common objective to make fusion a reality. *ITER,* 16 November. https://www.iter.org/newsline/-/2324.

Didier, C. (2010). Professional ethics without a profession: A French view of engineering ethics. In I. Van de Poel & D. Goldberg (Eds.), *Philosophy and engineering. An emerging agenda* (pp. 161–174). Berlin: Springer.

Doris, J. (2005). *Lack of character: Personality and moral behavior*. New York: Cambridge University Press.

Feinberg, J. (1999). Psychological egoism. In J. Feinberg & R. Shafer-Landau (Eds.), *Reason and responsibility*. (10th ed.). Belmont, CA: Wadsworth.

Frankel, M. (1989). Professional codes: Why, how, and with what impact? *Journal of Business. Ethics, 8*(2), 109–115.

Geertz, C. (1993). *Local knowledge*. London: Fontana.

Gupta, R. (2016). Why the road to nuclear fusion is necessarily bumpy. *The Wire*, 15 February http://thewire.in/2016/02/15/is-nuclear-fusion-worth-the-wait-21585/.

Hall, D., & Ames, R. (1987). *Thinking through confucius*. Albany, NY: State University of New York Press.

Harris, C., Pritchard, M., Rabins, M., James, R., & Englehardt, E. (2013). *Engineering ethics: Concepts and cases* (5th ed.). Boston, MA: Wadsworth Publishing.

Jamieson, D. (2001). *A companion to environmental philosophy*. Oxford: Blackwell.

Johnson, C. (2004). *Final Report: Review of the BFU Überlingen Accident Report*. 17 December. http://www.dcs.gla.ac.uk/~johnson/Eurocontrol/Ueberlingen/Ueberlingen_Final_Report.PDF.

Lucibella, M. (2015). Can the U.S. work well with international partners? *APS News*, June. https://www.aps.org/publications/apsnews/201506/partners.cfm.

Luegenbiehl, H. (2010). Ethical Principles for Engineers. In I. Van de Poel & D. Goldberg (Eds.), *Philosophy and engineering. An emerging agenda* (pp. 147–159). Berlin: Springer.

Luegenbiehl, H., & Fudano, J. (2005). Japanese perspectives. In C. Mitcham (Ed.), *Encyclopedia of science, technology, and ethics*. Detroit: Macmillan Reference.

Martin, M., & Schinzinger, R. (2010). *Introduction to engineering ethics* (2nd ed.). New York: McGraw Hill.

Mercer, M. (2001). In defence of weak psychological egoism. *Erkenntnis, 55*(2), 217–237.

Naess, A. (1973). The shallow and the deep, long-range ecology movement. *Inquiry, 16*(1–4), 95–100.

Nunes, A., & Laursen, T. (2004). *Identifying the factors that led to the Ueberlingen mid-air collision: Implications for overall system safety*. In: The 48th Annual Chapter Meeting of the Human Factors and Ergonomics Society. New Orleans, LA: USA.

Olivero, T. (2016). ITER fusion project: World leader in renewable energy. *Global Oil and Gas Industry News*, The OGM Energy Tech Culture ITER Fusion Project World Leader in Renewable Energy Comments. 22 February. http://theogm.com/2016/02/22/iter-fusion-project/.

O'Neill, J. (1992). The varieties of intrinsic value. *The Monist, 75*(2), 119–137.

Prince, M. (2004). Does active learning work? A review of the research. *Journal of Engineering Education, 93*(3), 223–231.

Rachels, J. (2003). Psychological egoism. In *The elements of moral philosophy*, (4th ed.). New York, NY: McGraw-Hill.

Rowlands, M. 2002. *Animals like us*. London: Verso.

Shaver, R. (2015). Ethical egoism. In E. N. Zalta (Ed.), *The stanford encyclopedia of philosophy (Spring 2015 Edition)*. Stanford: Stanford University. http://plato.stanford.edu/archives/spr2015/entries/egoism/.

Singer, P. (1975). *Animal liberation*. Nottingham: Old Hammond Press.

Slingerland, E. (2015). *Trying not to try: Ancient China, modern science, and the power of spontaneity*. New York: Broadway Books.

Smith, C. (2010). *What is a person? Rethinking Humanity, social life, and the moral good from the ground up*. Chicago: University of Chicago Press.

Strike, K., & Ternasky, P. (1993). *Ethics for professionals in education: Perspectives for preparation and practice*. New York: Teachers College Press.

The ITER Initiative. (2005). EUROfusion. 03 July. https://www.euro-fusion.org/2005/07/the-iter-initiative/.

Turney, R. (2007). The Uberlingen mid-air collision: Lessons for the management of control rooms in the process industries. *IChemE Symposium Series, 153* (pp. 1–5).

Van de Poel, I., & Royakkers, L. (2011). *Ethics, technology, and engineering: An introduction.* Malden: Wiley-Blackwell.

Wang, Q., & Zhu, Q. (2012). Traditional Chinese thinking and its influences on modern engineering and social development. In S. Christensen, C. Mitcham, B. Li, & Y. An (Eds.), *Engineering, development, and philosophy: American, Chinese, and European perspectives.* Dordrecht: Springer.

What Is ITER? ITER. https://www.iter.org/proj/inafewlines.

Whitbeck, C. (2011). *Ethics in engineering practice and research.* New York: Cambridge University Press.

Whorf, B. (1956). *Language, thought, and reality.* Cambridge: MIT Press.

Wilson, T. (2004). *Strangers to ourselves: Discovering the adaptive unconscious.* Cambridge: Harvard University Press.

Chapter 2

Working With Cases: The Importance of Concrete Learning

Chapter Objectives

Having read this chapter, answered the associated questions, and completed the included exercise, readers should be able to

- understand why studying cases is important to engineering ethics education, in general, and engineering ethics in global contexts, specifically;
- describe the process of and justifications for the steps involved in the case-study procedure;
- evidence the abilities involved in undertaking the initial steps involved in completing a case-study analysis.

2.1 CASE STUDIES

This second chapter introduces readers to the process of case-study analysis. As was indicated previously, studying cases is a useful tool for analyzing issues in applied ethics and recognizing complexities involved in making decisions with a real-world focus. This chapter further describes the benefits of and justifications for studying cases, outlines the steps involved in case-study analysis, and invites readers to practice completing this process with regard to a hypothetical case.

2.2 WHY STUDY CASES?

Chapter 1 introduced reasons for the use of cases in studying ethics. The benefits of doing so have been recognized over the last several decades, to the extent that the disciplines of engineering, professional, and applied ethics now focus on cases.[22] Studying cases helps both students and practitioners to

- learn actively, which has been demonstrated to increase understanding[23];
- determine the proper subject matter of ethical analyses;

---

22. For a discussion of the uses and centrality of case studies to these fields, see Beabout and Wennemann (1993), Delatte (1997), Luegenbiehl (1996), and Richards and Gorman (2004).
23. For more on this, again, see Prince (2004).

Global Engineering Ethics. http://dx.doi.org/10.1016/B978-0-12-811218-2.00002-3

- build an experience base regarding ethical issues before facing similar situations in the world;
- develop the ability to analyze and solve not only ethical but also engineering problems;
- recognize and understand the connection between the technical and ethical dimensions of engineering work;
- generalize by examining a variety of specific instances, thereby forming an integrated ethical perspective for oneself.

Although the study of cases is essential to an understanding of ethical issues, with an international approach to ethics, it is equally essential that engineering students and practitioners have a set of commonly agreed on set of guiding principles to which they can make reference, given their diverse backgrounds. The following discussion thus promotes the use of principles—to which the reader will be introduced shortly—in the case-study procedure.

2.3 STEPS FOR CASE ANALYSIS

2.3.1 Identifying Ethical Issues

In identifying ethical issues, it is important to keep in mind the definition of ethics from Chapter 1: *ethics concerns actions that have the potential to have a serious impact on the lives of others.* Thus, in identifying ethical issues while approaching a case, the primary concerns are those actions that have the potential to seriously impact the lives of others. These effects could be either direct or indirect, and the effects—or those affected—need not be explicit but, in many cases, should be deduced from the facts provided. In approaching a case, readers should carefully examine the facts of a situation in a general sense to see what information is provided, since, at first, there is the tendency to see only one prominent issue. In most cases, however, there will be more than one prominent issue.

Those new to ethical analysis often fail to recognize the presence of ethical issues in their lives, although doing so is one of the most crucial steps in ethical analysis.[24] When ethical issues are identified, they should be written out in question form, indicating that a conclusion should be reached, answering the question posed by the end of an analysis, for example, "should John have stolen/steal bread to feed his family?" Additionally, there are a variety of possible actors in any case—humans and nonhumans[25]—although cases may be presented from or about the perspective of only one individual:

---

24. Again, as a result of innate psychological biases, one of the biggest obstacles to acting ethically is the inability to discern the ethical dimensions of situations and recognize dilemmas as ethical ones. For more on this, see Bazerman and Tenbrunsel (2013).
25. For an account of both humans and nonhumans as "actors"—and discussions of the ethical and political ramifications of these interactions—see, for example, Latour (1992, 1994, 2004).

- Identify and formulate at least five ethical issues you have encountered in either your professional life as an engineer—internship or work environments—or personal life. (Remember that ethical issues are about actions that have the potential to seriously impact the lives of others—directly or indirectly—and should be written out in the form of a question, concerning what individuals or groups should or should not, might or might not have done.)
- Additionally, refer back to the case of the Überlingen midair collision, identifying and formulating at least five ethical issues present in that case.

2.3.2   Narrowing the Focus

Given the variety of issues that can be found in even a relatively short case, a full discussion of all the issues in any case would be difficult if not impossible. Here completeness in the analysis of a few issues is preferable to superficiality in an analysis of many, since the former progressively deepens one's understanding of ethics. Analysis should thus focus on only a couple of the main issues in a case. This will often involve the revision of initial questions, changing them so as to encompass subsidiary questions that have appeared. In this way, the complexities of a case can be retained, while the topics under consideration can be arranged in a hierarchical fashion, from the most to the least important:

- Referring back to the lists of ethical issues your wrote out before—concerning those you have encountered in your professional or personal life and identified in the Überlingen case—decide which two ethical issues from each list are the most important.
- Write out a brief justification for your decisions, why you feel these are the most important ethical issues.
- If necessary, rewrite these issues to encompass any subsidiary issues you think are important.

2.3.3   Determining Relevant Facts

When dealing with an ethical issue in real life, huge numbers of facts about situations could be listed. Facts related to case studies on engineering ethics could be grouped into the following three categories: (1) material facts—those concerning what went wrong/right from the perspective of material conditions and circumstances, which could be associated with the specifically engineering dimensions of incidents, for example, failures of the main and backup communication systems in the Überlingen midair collision; (2) facts regarding individuals—those concerning the action or inaction of individuals, for example, Peter Nielsen's failure to realize that 757 and 154 were flying at the same altitude; and (3) facts regarding organization—those concerning the ways that persons interact with persons, persons interact with technology, etc., in terms of business, government, etc., policies and guidelines, for example, the training of the Russian pilots to give priority to orders from air traffic control.

As facts may not always fit clearly into any one of these categories—overlapping between different ones and being connected through relations of implication and consequence to other facts—this categorization is meant to be heuristic in nature, assisting in the practice of identifying relevant facts. Additionally, although written cases restrict the reader to the facts mentioned, much of the material presented could still be superfluous in resolving the issues under consideration. Initially, one should make a list of all the facts relevant to the questions posed, grouping them in the above-described categories. This might require modification as the analysis proceeds. Next, facts that are not provided—but appear to be relevant to resolving the issues under consideration—should be listed as well:

- For each of the two most important ethical issues you identified above, list six facts that would help to answer these questions, two for each of the three categories: (1) material facts, (2) facts regarding individuals, and (3) facts regarding organizations.
- Additionally, for each question, list two missing facts that would help resolve the issues under consideration, facts concerning information missing from this case—to either you or the persons and/or organizations involved.

2.3.4   Making Reasonable Assumptions

In working with cases, students often complain they do not have enough information, using this as an excuse not to make decisions. Students might fail to recognize that this is also a common feature of real life, where reason is used as a supplement to make up for a lack of information. One needs to do the same in case-study analysis. It is possible to arrive at reasonable—although not certain—assumptions about purported facts through a process of inference. These assumptions add uncertainty in attempting to resolve ethical issues related to case study, but this uncertainty is preferable to the claim that no conclusion is possible. Making assumptions, however, is not an open-ended activity.

Assumptions should be justified, based on given data and a reasonable understanding of ordinary life. In addition, assumptions should be relevant to the particular issues under consideration. In Chapter 1, for example, the assumption was made that people tend to act for their own gain. This assumption is based on psychological studies and observations of general human behaviors. As information on the subject is limited, however, this would not be a fact regarding human motivations. Rather, the assumption is necessary as a background condition for the study of ethics:

- Referring back to the missing facts helpful to resolving the two most important ethical issues you identified above—in your professional/personal life and the Überlingen case—what reasonable assumptions can you make? Explain your reasons for making these assumptions.

• Have there been other times in your professional or personal life when you have had to make important decisions but lacked information to make these decisions? List them. What did you do? What assumptions did you make in attempting to arrive at a decision?

2.3.5   Undertaking Definitional Clarification

One of the central tasks of philosophical analysis is gaining clarity regarding concepts. Without such clarification—which is essentially a definitional activity—communication between different persons and people can fail. This is important when dealing with issues in international contexts, where cultural connotations associated with words differ significantly.[26] In addressing an ethical issue to be resolved, it is thus important to explain the ways any central concepts are used, especially those subject to differing interpretations. This is especially true with terms that have "value connotations," implicit meanings regarding understandings of right and wrong, good and bad, etc.:

• Returning to the previous exercises, have you used terms or concepts that could be considered ambiguous or "conceptually vague"—lacking one clear definition or using terms/concepts in ways that could be confusing to others? If so, then list them.
• Imagine you're someone else and then write out how someone else might understand each of these terms/concepts.
• Provide definitional clarification regarding how you meant each term/concept to be used/how you intend others to understand them.

2.3.6   Conducting Ethical Analysis

"Ethical analysis" here means the application of principles further explained and justified throughout this text.[27] In making ethical judgments, alternatives to this approach also exist, some of which will also be discussed. At present, it is sufficient to note that this process of applying principles is crucial to making ethical judgments and that, at times, principles may be insufficient. This means that in the process of application, a simultaneous process of refining and justifying the prior foundations for making ethical judgments should also occur.[28] In many ways, this

---

26. For a discussion of the ways this interferes with accurately assessing cross-cultural values, see Kulich and Zhang (2010).
27. Chapters 3–11 list and justify these principles. For a full list of these principles, see Appendix I.
28. This process is similar in nature to "reflective equilibrium" in Rawls (1999), "principlism" in the field of biomedical ethics by Beauchamp and Childress (2008), and the "ethical cycle" as proposed by engineering ethicists Van de Poehl and Royakkers (2007, 2011). Regarding the nature of "reflective principlism," a modification of principlism appropriate to teaching engineering ethics, see Kisselburgh, Zoltowski, Beever, Hess, and Iliadis (2014).

process should be similar in nature to the one that occurs in the common-law tradition, in the interplay between laws and their interpretations in courts.

Principles

Cases

It is important to remember that no externally provided set of principles can ever serve as a final authority. Individuals must exercise their reason in making judgments, for they are the ones ultimately accountable for their decisions.

The first step in analysis consists in reviewing the set of available principles and deciding which of them applies to the issues under consideration. For example, if someone takes bread without paying, then this action could be considered ethically wrong according to the principle "one should not steal." In applying principles to specific cases, however, conflicts may well arise between principles. The next step then is to decide whether such conflicts exist. If only one principle applies, then no conflict will exist, although this is seldom the case. If conflicts exist, then it must be determined whether one principle should override another. This may lead to a general prioritization of one principle over another, but the established hierarchy might also only apply to the case at hand. For example, if someone takes bread to feed his or her starving family, then according to the principle "one should feed his or her family," a conflict of principles arises.

Deciding which principles should take precedence—either in general or with reference to a specific case—is based on the ability to reason in relation to the given set of facts. The final step in analysis consists in deciding whether the set of principles available is sufficient to reach decisions regarding the case— to resolve the ethical issues—or whether additional principles for action are needed. Principles concerning general and personal ethics might be especially relevant in resolving ethical issues:

- Which ethical principles—rules for behavior regarding right and wrong—are important in your personal and working, professional life? List them.
- Are these principles the same in your personal and working, professional life? Why or why not?
- Which of these principles apply to the ethical issues you previously raised concerning your professional/personal life and the Überlingen case—in other words, which of these principles helped/would help you resolve the ethical issues/answer the questions you posed above?
- Do any of the principles conflict in the two situations? If so, then how would you "hierarchize"—rank in priority—these principles? Is this hierarchy generally applicable or specific to the case under consideration? Why?
- Are there any other ethical principles that might be applicable to these issues—rules for how people should conduct themselves regarding right and wrong that would help resolve the ethical issues/answer the questions you posed before?

2.3.7   Reviewing the Process

A complete case analysis is an "iterative process," which means that as case study proceeds, new issues might be recognized, additional facts may be needed, or confused concepts could require clarification. Going back and revising materials from earlier stages of analysis causes no harm. Aiming at completeness is more important than aiming at linearity and straightforwardness. If this step is included, a written analysis will more closely mirror real-life approaches to dealing with ethical problems.

The other aspect of process review is determining whether mistakes have been made throughout the analysis. Here it is important to make sure that consistency and objectivity have been maintained in the process. Just as in solving engineering problems, the ability to use reason is the key instrument in ethical analysis.[29] Reason can, of course, be influenced by subjective biases and subconscious elements. It is important that one be aware of and guard against such biases and elements, so as to reach a conclusion that has the potential to convince others of its reasonableness:

- Complete the process review on the work you've done so far concerning the ethical issues you raised about your professional/personal life and the Überlingen case—review the first 6 steps of the case-study procedure one more time. Have you missed any important facts? Do any of the concepts you have used require further clarification? Have you maintained consistency and objectivity throughout? Have your own unconscious or subjective biases influenced your analyses?
- When dealing with an ethical situation, do you think it is acceptable that your reasoning be influenced by biases? Why or why not? Explain a situation when biases might and might not be tolerable.

2.3.8   Resolving the Issue

After the process review, answers to the initially posed questions can be stated with some certainty. In real life, decisions have to be made. Even a failure to act typically results in consequences. In an analysis of the type conducted here, however, some claim they are unable to arrive at a conclusion, either because they do not have sufficient information or because they are unable to decide on the appropriate application of principles. Because of the nature of real-life consequences, however, such avoidance is acceptable only in very rare instances.

This step then involves clearly stating a conclusion. In addition, a short justification should be provided as to why the conclusion is appropriate. After working

---

29. Again, for more on this analogy, see Whitbeck (2011), and for a fuller discussion of this position, see Van de Poehl and Royakkers (2011).

on a case, a certain conclusion might appear to follow so obviously that no further explanation is required. However, the reasons for stated conclusions might not be as obvious to those unfamiliar with the case under consideration:

- Resolve the ethical issues you identified as the most important with regard to your professional/personal life and the Überlingen case, answering the questions you posed above.
- Provide brief justifications for these responses.

2.3.9   Identifying Practical Constraints

From a traditional ethical perspective, the previous step would complete an analysis. As indicated before, however, the approach taken here aims at recognizing and coming to terms with the realities of life in case-study analysis. For this reason, it is important to examine whether any practical constraints exist that would make an ethically acceptable action too difficult to expect from someone in the set of circumstances under consideration. This could include, for instance, deeply ingrained cultural norms. In the literature, these kinds of reasons for not acting ethically are called "excusing" or "mitigating conditions," and they would absolve someone for a part or all of the responsibility for not acting ethically. Rather than conceiving this inaction as simply a failure to adhere to ethical standards, it is more appropriate to think of these excusing conditions as part of a fuller and more complete ethical analysis. Practical constraints are a fact of life and should be acknowledged as such. Recognizing such constraints can allow one to avoid ethical problems in the first place:

- In your personal and/or professional life, have you acted in an unethical manner? List the practical constraints that prevented you from acting in an ethical fashion, explaining the nature of these constraints.
- Referring to the answers you gave to the ethical issues in your professional/ personal life and the Überlingen case, do practical constraints exist that would prevent a resolution in the way you described? If so, then list and explain the nature of these constraints in relation to the issues.
- Give an example of how deeply ingrained cultural norms can create mitigating conditions in relation to the resolution of ethical issues.

2.3.10   Avoiding Ethical Problems

In addition to addressing the ethical issues with which one will be confronted, case-study analysis helps one to think ahead and avoid serious ethical quandaries in the first place. The most successful individuals are those who think ahead and plan for potential consequences of their actions. The best chess and basketball players, for example, are those who anticipate their opponents' moves ahead of time, perhaps even before the opponents themselves. This requires an understanding of one's opponents specifically and human psychology in general.[30] The same principles apply in dealing with ethical issues.

Although circumstances sometimes require having to face and make difficult decisions with regard to ethical problems, at other times such problems can be avoided in the first place.[31] While this is easier to determine in retrospect than at the time—the saying in English, "Hindsight is 20-20," refers to this phenomenon—even hypothetical discussions can help processes of decision-making. However, this step should come at the end of case-study analysis rather than at the beginning, so as not to preclude the possibility of making hard choices:

- With regard to difficult ethical decisions you have made in your own life, might you have avoided making such decisions had you acted differently earlier? In other words, had you made different decisions previously, might you have avoided being in a position where you were confronted with a difficult ethical decision?
- Referring again to the questions you raised regarding your professional/personal life and the Überlingen case, could these issues have been avoided if different decisions had been made earlier? Explain your answer.

CASE STUDY—A HYPOTHETICAL CASE FOR STUDENT ANALYSIS: A HEATING UNIT DEFECT

The following is a hypothetical case included for readers to conduct their own sample analysis, either individually or in small groups. Begin by reading the case included below and, after having done so, write out the following:

1. Five ethical issues that appear in the case, stated as questions regarding what should or should not—might or might not—have been done.
2. State what you consider to be the main ethical issue, briefly giving reasons for why you consider it to be the most important ethical issue.
3. List facts provided in the case that are relevant to resolving the main ethical issue/answering this question, categorizing these in terms of material-/engineering-related facts, facts concerning individuals, and facts regarding organizations. Additionally, list any missing facts—information that is absent but would be relevant to resolving this issue.
4. List reasonable assumptions you can make regarding the missing facts/absent information needed to resolve the main ethical issue you identified above.

---

30. Regarding the benefits involved in empathizing with others and visualizing one's intended conduct, see Lu, Dane, and Gellman (2005).
31. The philosopher John Doris claims this is a major takeaway from findings related to situationist psychology: since small environmental factors make big differences in the ways persons and people behave—especially unconsciously—rather than "character," to avoid acting unethically, persons and people should avoid certain types of environments in the first place (Doris, 2005).

5. Review your work thus far, undertaking any necessary conceptual clarification.
6. Identify ethical principles relevant to the issue under consideration, determining a hierarchy and thereby resolving possible conflicts between them. (As readers have not yet been introduced to the full list of and justifications for principles of global engineering ethics—outlined in Chapters 3–11—they should refer to principles of right action that seem intuitive or with which they are already familiar, for example, "don't hurt others" or "do onto others as you would have others do onto you."[32])
7. Review your work thus far, determining the need for adjustments to any of the prior steps.
8. Resolve the ethical issue under consideration, reaching a final answer to the main question you identified previously. Briefly justify your decision.
9. Identify any practical constraints that might influence a purely ethical judgment.
10. Discuss ways the main ethical issue could have been avoided had different actions been taken earlier. Identify at least three.

John Smith is a young, single engineer who graduated from college about 2 years ago. Since that time, he has worked for the Kastor Manufacturing Company. His most recent assignment, which began 4 months ago, consists in developing a part to be integrated into a new electric heating unit the company is planning to market, mainly to young people living in small apartments. Smith is happy in his work since he likes and respects his coworkers. Smith's home life is a bit more troublesome: he lives with his elderly parents who largely depend on him to take care of them, since his father is unable to work due to a debilitating disease, and his mother is depressed all the time because of this.

Smith's work on the project goes according to schedule, although he is under some pressure to complete his part of the project. One day, however, his tests show that the part on which he is working will fail within the first 2000 h of service in approximately one out of every one thousand units in which it will be installed. Failure of the part could potentially result in the unit's overheating and a fire. Further testing reveals to Smith that in order to correct this defect, he will basically need to start his work over again. Naturally upset by this finding, Smith reports this problem to his supervisor, John Brady.

As usual, Brady is sympathetic to Smith's concerns, but when Smith asks for extra time to complete his portion of the project, Brady tells him in no uncertain terms that the part must be ready on time. Management has told Brady that the heater must be available on time, since Kastor's primary competitor will release a rival heating unit approximately a month after Kastor is

---

32. Again, a full list of the principles of global engineering ethics can be found in Appendix I and an analysis of the hypothetical "Case of Curious George" in Appendix II.

scheduled to release its own. Brady also tells Smith, confidentially, that Kastor is facing financial difficulties; at present, this information should be kept from the employees.

When Smith continues to express his concerns, Brady assures him that he will personally see to it that corrective actions are taken later. Brady will insist to management—at the possible expense of his job—that the part for which Smith is responsible be redesigned and any units that have already been sold be recalled. He tells Smith that in terms of the big picture, this is the most cost-effective way to handle the situation. Plus, Brady tells Smith, it is likely that most of the units will be recalled before any problems occur, since the units are likely to be used less than 2000 h during an entire heating season, and the recall should occur within 6 months.

EXERCISE—ANALYZE ENGINEERING A BETTER GLOBAL FUTURE

The case of the Überlingen midair collision involves the actual loss of life, and that of a heating unit defect concerns potential breaches in public safety. To ensure public safety, the case-study procedure can be used on incidences such as these to determine better courses of action, ones that keep people safe. By contrast, the case of Engineering a Better Global Future: Fusion Power Across Borders involves neither the actual loss of life nor potential breaches in public safety. Rather, it could be considered a case of "aspirational ethics," one where individuals and organizations have behaved in more exemplary manners and from which lessons can be learned. To ensure better outcomes, the case-study procedure can be used on situations such as these, not only keeping people safe but also improving their lives:

- Returning to Engineering a Better Global Future in Chapter 1, complete the same 10 steps of the case-study procedure you carried out on a situation related to your professional/personal life and the Überlingen case.
- Other than case studies, what do you think are ways engineers in global contexts can better understand ethical issues? Explain your answer. Have you participated in these activities? Why or why not?

2.4 SUMMARY

In combination with the use of reason and role responsibilities of engineers, case-study analysis is an essential aspect of a global approach to engineering ethics. The case-study procedure is part of a "ground-up," reciprocal approach to ethics—examining individual instances following the case-study procedure, generalizing an ethical framework on this basis, bringing this framework to bear on further instances, and so on and so forth. This approach is important as general disagreement exists regarding ethical theories, especially in global and cross-cultural contexts, and because the fields of engineering and technology are constantly and quickly changing, such that a malleable and adaptive approach to ethics is necessary. As the hypothetical case of A Heating Unit Defect makes clear, in many instances, it is difficult to determine a clear and correct course of ethical action. Rather, one has to make reasonable assumptions when information is unavailable, determining and employing relevant principles to the best of one's ability.

REVIEW QUESTIONS

1. List three skills utilized by students and practitioners during case-study analysis. Explain two situations in which these abilities would be essential for a global engineer.
2. Explain "value connotations."
3. What makes definitional clarification especially important in relation to the global context of engineering ethics?
4. What steps should one take if conflicts arise between available principles during ethical analysis?
5. Why is "iteration" important to the case-study procedure?

REFERENCES

Bazerman, M., & Tenbrunsel, A. (2013). *Blind spots: Why we fail to do what's right and what to do about it*. Princeton: Princeton University Press.

Beabout, G., & Wennemann, D. (1993). *Applied professional ethics: A developmental approach for use with case studies*. Lanham: University Press of America.

Beauchamp, T., & Childress, J. (2008). *Principles of biomedical ethics* (6th ed.). New York: Oxford University Press.

Delatte, N. (1997). Failure case studies and ethics in engineering mechanics courses. *Journal of Professional Issues in Engineering Education and Practice, 123*(3), 111–116.

Doris, J. (2005). *Lack of character: Personality and moral behavior*. New York: Cambridge University Press.

Kisselburgh, L., Zoltowski, C., Beever, J., Hess, J., & Iliadis, A. (2014). Effectively engaging engineers in ethical reasoning about emerging technologies: A cyber-enabled framework of scaffolded, integrated, and reflexive analysis of cases. *In: 121st ASEE annual conference and exhibition. Indianapolis*.

Kulich, S., & Zhang, R. (2010). The multiple frames of "Chinese" values: From tradition to modernity and beyond. In M. Bond (Ed.), *The Oxford handbook of Chinese psychology*. New York: Oxford University Press.

Latour, B. (1992). Where are the missing masses? In W. Bijker & J. Law (Eds.), *The sociology of a few mundane artifacts: Shaping technology/building society*. Cambridge: MIT Press.

Latour, B. (1994). On technical mediation—Philosophy, sociology, genealogy. *Common Knowledge, 3*, 29–64.

Latour, B. (2004). *Politics of nature: How to bring the sciences to democracy*. C. Porter (trans.) Cambridge: Harvard University Press.

Lu, Y., Dane, B., & Gellman, A. (2005). An experiential model: Teaching empathy and cultural sensitivity. *Journal of Teaching in Social Work, 25*(3-4), 89–103.

Luegenbiehl, H. (1996). Ethical analysis using case studies. In T. Kent & M. Gentry (Eds.), *The practice and theory of ethics*. Indianapolis: University of Indianapolis Press.

Prince, M. (2004). Does active learning work? A review of the research. *Journal of Engineering Education*, *93*(3), 223–231.

Rawls, J. (1999). *A theory of justice* (2nd ed.). Cambridge: Harvard University Press.

Richards, L., & Gorman, M. (2004). *Using case studies to teach engineering design and ethic*. In: Proceedings, American society for engineering education conference.

Van de Poehl, I., & Royakkers, L. (2007). The ethical cycle. *Journal of Buisness Ethics*, *71*(1), 1–13.

Van de Poehl, I., & Royakkers, L. (2011). *Ethics, technology, and engineering: An introduction*. Hoboken, NJ: Wiley-Blackwell.

Whitbeck, C. (2011). *Ethics in engineering practice and research*. New York: Cambridge University Press.

Chapter 3

Engineering Professionalism and Professional Organizations

Chapter Objectives

Having read this chapter, completed the included exercises, and answered the associated questions, readers should be able to

- describe the ways "profession" has been defined, the relationship of professions to society and individual professionals, and the "contract-model" account of professions;
- give reasons engineering should and should not be considered a profession, in relation to the history of engineering and current state of its professional organizations;
- explain the nature of and reasons for codes of ethics for engineers, with reference to both the history of codes of ethics and the American Society for Mechanical Engineering (ASME) code of ethics as a sample.

CASE STUDY ONE—MCDONNELL AND MILLER, AND THE ASME: PROFESSIONALISM IN QUESTION?

On May 17, 1982, the Supreme Court, in a 6–3 ruling, found the American Society for Mechanical Engineering (ASME) guilty of having violated the Sherman Antitrust Act, which aims at preventing unfair and collusionary business practices. The ASME was eventually ordered to pay the Hydrolevel Corporation (henceforth Hydrolevel) $4,750,000. This ruling was a result of actions taken by members of the ASME that unfairly disadvantaged Hydrolevel, which produced boiler cutoff valves.

This incident called into question the integrity of the ASME and its professional members specifically and calls into question the potential integrity of professional organizations and their group members in general: it illustrates conflicts of interests that can arise between corporations and professional organizations, between the role responsibilities engineers have in their business and professional lives. This case can serve as a starting point to examine the nature and roles of professionals and professional organizations within society in general and engineering specifically.

In 1971, Hydrolevel received a large order for boiler cutoff valves from the Brooklyn Gas Company. At the time, Hydrolevel was a relatively small company with a minor share of the boiler cutoff valve market. Until then, the Brooklyn Gas Company had been a major client of the much larger McDonnell and Miller Incorporated (hereafter MM). Disturbed by this situation, Eugene Mitchell, MM's vice president for sales, approached John James, MM's vice president for research, regarding a course of action. At the time, James was not only the vice president for research at MM but also a member of the ASME's heating boiler subcommittee (henceforth the ASME's HBS). In hopes of defaming Hydrolevel and gaining back market shares for MM, James and Mitchell approached T. R. Hardin, the chairman of the ASME's HBS. The three men met over dinner and discussed an interpretation of the ASME's Boiler and Pressure Vessel Code (henceforth the B-PV).

The B-PV dictates the safe design and operations of boiler valves. As boilers can explode if water levels become too low, the B-PV stated that each "automatically fired steam or vapor system boiler shall have an automatic low-water fuel cutoff, so located as to automatically cut off the fuel supply when the surface of the water falls to the lowest visible part of the water gauge glass" (Harris, Pritchard, & Rabins, 2009). Whereas the boiler cutoff valves manufactured by MM would shut off a boiler immediately if water levels became too low, those produced by Hydrolevel were time delayed.

James and Mitchell brought this code and these facts to Hardin's attention in the hopes that—as the chair of the ASME's HBS—Hardin would deem the cutoff valves produced by Hydrolevel unsafe. This would result, in turn, in MM gaining back market shares from Hydrolevel. Hardin agreed to James and Mitchell's interpretation of the code. In addition to acting as the chair of the ASME's HBS, Hardin was the vice president of the Hartford Steam Boiler Inspection and Insurance Company (hereafter HSBIIC). Both James and Hardin thus held multiple roles—ones in companies that operate for the sake of profits and ones in the ASME that operates for the sake of engineering professionalism. In this situation, their responsibilities came into conflict: they abused their positions in the ASME for the sake of profit.

With this agreement in place, James sent a letter to the ASME's HBS, including Hardin's opinion. When Bradford Hoyt, the secretary of the Boiler and Pressure Vessel Committee, received James' letter, he forwarded it to the appropriate subcommittee chair, Hardin. Hardin then responded in agreement with James' letter, without either consulting or obtaining the approval of the other subcommittee members. As a result of these actions, Hydrolevel suffered financial losses: the company eventually went bankrupt.

Salespersons at MM used James' letter and Hardin's response as a basis for claiming Hydrolevel's valves were unsafe and should not be purchased, which Hydrolevel learned of from a customer in early 1972. In Mar. of that year, Hydrolevel requested that the ASME review and correct its assessment of their valves. The ASME's HBS met in Jun. and upheld part of Hardin's original interpretation of the B-PV. In Aug. of 1975, Hydrolevel sued the ASME, HSBIIC, and MM. The HSBIIC and MM both agreed to out-of-court settlements with

Hydrolevel, although the ASME did not. The ASME argued that it was not liable as an organization, since the actions that harmed Hydrolevel were those of individual members of the ASME rather than the ASME itself.

The case went to court, and in Feb. of 1979, the ASME was ordered to pay Hydrolevel $7,500,000, although this amount was subsequently reduced. The decisions of the courts were based on the tremendous power of the ASME as a professional organization: the ASME has the ability to issue decisions on design, operation, guidelines, and procedures that affect the lives and businesses of millions of people. As a result of this case, the ASME changed a number of operating and review procedures to protect against perceived, potential, and actual conflicts of interest.[33]

Although this case is generally analyzed in the context of perceived, potential, and actual conflicts of interests—to which this text returns in Chapter 9—it can also serve as a starting point to examine the nature of professionalism, both individually and organizationally. Although most would agree that James and Hardin abused their positions in the ASME, the responsibilities that should have followed from their roles as professionals within a professional organization (the ASME) might not be clear.

What does professionalism mean? What kinds of relations exist between professions and society? Which criteria should be used to determine what counts as a profession? Why should society care about professions and—just as importantly—why should professions care about society? Most importantly, what is the relation between professionalism and ethics? As answers to these questions are not simple and—in many cases—widespread misconceptions exist regarding the nature of professionalism, professionals, professional organizations, and their relations to society, those are some of the questions this chapter addresses.

EXERCISE—MCDONNELL AND MILLER, AND THE ASME

Before moving on to the topic of professionalism, complete the case-study procedure with regard to the case of McDonnell and Miller, and the ASME. To refresh your memory, these steps are as follows:

1. Identify ethical issues—list five ethical issues.
2. Narrow your focus—narrow your focus to one ethical issue.
3. Determine relevant facts—list facts that would be relevant to resolving this ethical issue and any missing information.
4. Make reasonable assumptions—make reasonable assumptions regarding missing facts.
5. Undertake definitional clarification—clarify any terms/ideas that might be unclear, in relation to both issues and facts, paying special attention to those that have "value connotations."

---

33. This account is additionally based on Davis, Jones, and Wells (1986), Harris et al. (2009), and Department of Philosophy and Department of Mechanical Engineering (n.d.).

6. Conduct ethical analysis—list any ethical principles that would be relevant to resolving the ethical issue you identified, noting and prioritizing possibly conflicting principles.
7. Review the process—go back through the previous steps again, checking your work.
8. Resolve the issue—answer the question you posed before, giving a brief justification for your answer.
9. Identify practical constraints—do practical constraints exist that could excuse either individuals or organizations from the answer given? If so, then list these conditions.
10. Avoid ethical problems—how might the ethical problems you listed have been avoided in the first place?

3.1 "PROFESSION": MORE THAN YOU MIGHT THINK

This chapter explores the notion of "profession" and related concepts of "professional" and "professionalism." This represents a traditionally US manner of conceiving ethics within the field of engineering. Although this approach has developed in the United States, as engineering becomes an increasingly global occupational discipline, it will standardize along professional lines, developing in terms of the professional model.[34]

An often initially given—but ultimately incorrect—definition of a "professional" is someone who is paid to do something, such as a "professional" baseball player. This definition is incorrect, since persons in nonprofessional occupations are paid as well. Someone engaged in selling admission tickets at a ballpark, for example, would not be considered a professional. In fact, one of the characteristics of most professions is that its members are expected to do some type of *pro bono* work, work for free, for the benefit of either the professional community or the community as a whole. The idea of a profession will, therefore, be more complicated than initial, common conceptions might indicate.

3.2 THREE WAYS OF DEFINING "PROFESSION"

Several proposals have been made to define "profession." Definition by:

1. Paradigm or ideal type—This method focuses on identifying a group widely recognized as a profession, using the characteristics of this group to evaluate others, thus creating a hierarchy of professions and semiprofessions, based on how closely they resemble the paradigm. Medicine is the most commonly

---

34. The nature and development of professionalism and professional organizations in China and India are addressed at the end of this chapter.

used, and its characteristics consist in a theoretical knowledge base, long period of training, specialized skill set, set of licensing requirements, rights to and expectations of autonomous action, codes of ethics, and professional associations.

2. Central characteristic—This method attempts to arrive at the essence of what it means to be a profession, most commonly located in the "service ideal." This refers to the establishment of a profession when the public recognizes that a specialized, highly skilled, and necessary activity can be performed best if a small group is delegated responsibility as the sole provider of that service. This perspective thus focuses on conditions such as the exclusive right to perform a particular service, autonomous judgment of the quality of the professional activity, and control of entrance requirements to the profession. For instance, a task assigned to medicine is control over medical services, to law that over legal services, and so on. Individuals who are not members of these professions are not allowed to provide these services.

3. Definition by prestige—This method begins with the actual circumstances of occupational groups, rather than ideals or essences. It assumes that occupational groups want the prestige and power associated with being a profession, and that, based on self-interests, they will take the steps necessary to achieve that title: occupational groups will attempt to acquire whichever characteristics society demands from the group it refers to as a "profession," for example, the formation of professional associations and the development of codes of behaviors. Groups identified as professions are, therefore, simply those that are called "professions" by society. Based on this definition, any occupational group could be a profession.[35]

Readers should note that all three of these approaches emphasize the idea of professional ethics: the first because, going back to the Hippocratic oath, medicine has stressed ethics; the second because it can demonstrate to society that a profession is fulfilling its end of a social contract; and the third because it can serve as a way to create a favorable public impression.

All three of the above definitions have elements of truth: first, groups are modeled on ideal professions. Historically, there have been three of these: medicine, law, and the clergy, which traditionally served a teaching function. Second, society perceives professional groups as performing particular, necessary services, and society rewards professional groups accordingly. Third, groups do seek out the prestige of "profession" status in ever-increasing numbers. In thus analyzing the idea of "profession," it is valuable to keep all three perspectives in mind. All three views can be synthesized in what could be called the "contract model."

---

35. For more on the natures and developments of professions and professional organizations, see, for example, Bucher and Stelling (1969), Bucher and Strauss (1961), and Larson and Larson (1979).

3.3 THE RELATIONSHIP OF PROFESSIONS TO SOCIETY: THE CONTRACT MODEL

In theory, professions need not exist: society chooses groups to perform particularly valuable services. Medicine, for example, has long been recognized as such a service. A relationship of exchange is thus established between society and a particular professional group; a trade is made. The professional group performs a service for society in an exemplary fashion and, in exchange, society gives the professional group a monopoly on the service, prestige, and (generally) good pay. Both society and professions have generally perceived this arrangement as a fair trade. At times, however, such a relationship can be seen breaking down.

Either society no longer needs the services performed by the professional group, or the professional group no longer fulfills its part of the social contract—in other words, carrying out the service ideal—instead appearing to be primarily self-interested. This can even occur with the core professions, as is reflected in the sinking reputation of lawyers in the United States.[36] A lessening of prestige would thus occur, and society could potentially withdraw the privilege of an occupational groups' status as a profession. In the meantime, other groups strive to obtain professional status, resulting in a hierarchy of professions that changes over time, based on the prestige accorded to occupational groups by society. As society places less trust in lawyers, for example, "mediators" have striven to provide some of the services traditionally associated with the legal profession at a higher quality and lower costs.[37]

3.4 CHARACTERISTICS OF A PROFESSION

Based on the above description, certain attributes can be used to characterize a profession, not all of which need be met, but that establish a hierarchy among professions:

1. The development of a specialized body of knowledge, based on a theoretical framework.
2. Learning a set of skills that puts this knowledge into action—in other words, professionals ultimately deal with applied knowledge.
3. A long period of formal education necessary to acquire the above described knowledge and skills and to socialize prospective members into the profession.
4. The profession controls educational and extraeducational requirements for admission into the profession. This often takes the form of accrediting educational programs and licensing individual professionals.

---

36. Concerning the sinking reputation of the legal industry, see, for example, Re (1994), Galanter (1993), and Galanter (1997).
37. For more on this, see Hensler (2003).

5. The profession controls the actions of individual professionals—in other words, a way of punishing the actions of individuals who fail to adhere to professional norms.

6. Lifelong membership in the profession: once someone becomes a member of the profession, unless that person is formally evicted, he or she remains a member of the profession for life, even if and while undertaking different activities. This attribute characterizes "professional identification."

7. The existence of a professional culture that establishes the norms and behavioral patterns appropriate for members of the profession.

8. The profession fosters individual autonomy and authority. The professional is an independent practitioner who serves clients rather than an employee who has a boss. Based on the knowledge and skills of a professional, the professional has authority over the client, even though the client pays.[38]

9. A way of demonstrating to society that the service ideal is met: this is generally achieved through principles established in codes of ethics, to which we return below.

10. The establishment of professional organizations, to disseminate technical knowledge and enhance the professional culture.

11. Society accords prestige and good pay to the members of the profession.

3.5 THE RELATIONSHIP OF A PROFESSION TO THE INDIVIDUAL PROFESSIONAL

In general, the profession serves as an intermediary between society and individual professionals. Society grants the profession an exclusive right to provide a service—a monopoly—as no other group is in a better position to determine whether the service provided is being performed adequately. This implies that the profession should establish control over the actions of its members. As mentioned above, professions do so by establishing criteria for education and membership and by punishing the wrongdoings of members. These conditions are often given legal status in the United States. Additionally, however, professions also attempt to foster a sense of individual responsibility in their members, through the development of professional autonomy, and a sense of duty to the professional community and society as a whole.

---

38. This relation would generally be characterized as one of "paternalism." This word is based on the Latin root *pater*, which means "father." Hence, a professional would relate to a client as a father does to a child, knowing more and making decisions for a child, although ultimately holding the child's best interests as a top priority. Potential problems exist with this relation, topics related to which are covered in more detail in Chapter 8 on the notion of (professional) autonomy.

Professions (professional organizations)

Professions to society:
ensure service quality

Contract model

Society to professions:
exclusive service right

Professions to professionals:
regulate service behavior

Professionals to professions:
loyalty

Society → Opportunities, pay, and prestige → **Professionals**

← Service and loyalty ←

At times, professions in the United States have emphasized one or more of these characteristics too much: they have been accused of restricting membership for the economic gain of present members. They have been accused of punishing those who speak out against misdeeds in the profession. They have been accused of failing to punish the misdeeds of members. They have been accused of fostering an atmosphere where upholding the good of the profession becomes more important than the good of society. While there are instances of all of these, it is also important to remember that society continues to accord a great deal of respect to professions and the work of professionals[39]:

- In the case of McDonnell and Miller, and the ASME, list and explain some of the responsibilities the ASME failed to fulfill as a professional organization.
- Do you think professional organizations should be held accountable for the actions of their members? Why or why not?

3.5.1   Engineering as a Profession

Each profession benefits society in terms of its particular mission. The major mission of engineering consists in designing and being responsible for the production of technological devices. In contemporary society, this is an incredibly important and specialized function. However, in the United States at least, the professionalization of engineering has been a relatively recent phenomenon.

In terms of its historical tradition, engineering was primarily craft-based, emphasizing the "apprenticeship" rather than professional model. Schools of engineering did not develop until the mid-19th century, and even then these were organized on a shop floor model. In other words, their main emphasis was on *doing* engineering rather than *theoretical* knowledge. Thus, both educationally and occupationally, the professional model developed

---

39. Again, with regard to the natures and developments of professions and professional organizations, see Bucher and Strauss (1961) and Bucher and Stelling (1969).

quite slowly. Not until the last part of the 19th century were national professional organizations for engineers founded, and not until the early part of the 20th century were codes of ethics developed. Since that time, there has been intense activity to achieve and maintain the professional status of engineering.[40]

The present state of affairs is somewhat mixed: engineers are generally recognized by society as professionals, but their prestige is not as great as that of traditional professionals, such as doctors. Engineering has been termed "the invisible profession," since its members have generally not had the authority and autonomy associated with other professions.[41] Engineering has had difficulties establishing universal licensing requirements and controls over admission to the profession. Corporations can still hire students without engineering degrees as "engineers," and engineering has been unsuccessful in establishing more than a 4-year degree as the minimum requirement for admission to the profession.[42] However, at least with regard to public perception of engineering, as noted previously, the profession has been quite successful.

3.6 PROFESSIONAL ORGANIZATIONS

The profession plays an important role in the lives of individual engineers. It mainly does so through a variety of professional organizations. The following section considers the role of these organizations in more detail, by examining how engineering organizations have influenced the development of a moral basis for engineering through the establishment of codes of ethics.

3.6.1 Professional Organizations and Codes of Ethics: Some Examples

Based on the contractual model of professions discussed above, one of the most important functions professions exercise is control over their members: the profession guarantees to society the adequacy of work performed by the profession, and it must have the means to ensure this guarantee is met. Professional associations are generally responsible for exercising these controls over their members. As was shown in the case at the beginning of this chapter, professional associations, such as the ASME, are responsible for the development of technical standards, setting up of behavioral norms, and, additionally, the establishment of licensing and educational requirements.

Some professions have one central organization that exercises all of these functions, for example, the American Medical Association, and various specialty organizations devoted to subdisciplines, for example, the American

---

40. For more on the nature and development of engineering as a profession, see Layton (1986).
41. Regarding engineering as "the invisible profession," see Fitzgerald (2005).
42. For more on licensing requirements of professional engineers in the United States, see the NSPE's (National Society of Professional Engineers) website: http://www.nspe.org.

Psychiatric Association.[43] In the United States, engineering has no such central organization, probably in large part because engineering lacks universal licensing requirements. Instead, different organizations administer the various functions of the profession:

1. *National Society of Professional Engineers* (NSPE)—responsible for the development of licensing procedures. However, in the United States, licenses themselves are granted by individual states through the administration of individual state professional engineering organizations. In the United States, less than 20% of practicing engineers are licensed, due to an "industrial exemption." This refers to the fact that, in companies, licensed engineers can certify (approve) the work of other, nonlicensed engineers. As a group, civil engineers are the most heavily registered (licensed).[44]

2. Technical organizations—each major branch of engineering has its own organization responsible for the transmission of new knowledge and fraternal bonding. Many smaller, subspecialty organizations also exist. The major technical organizations include the *American Society of Mechanical Engineers* (ASME), *American Society of Civil Engineers* (ASCE), *American Institute of Chemical Engineers* (AIChE), and *Institute of Electrical and Electronics Engineers* (IEEE). These organizations are responsible for the development of technical standards within engineering. The *American Association of Engineering Societies* (AAES) was formed to bring together the various professional organizations in a common forum, but its effectiveness has been somewhat limited.

3. *Accreditation Board for Engineering and Technology* (ABET)—responsible for accrediting engineering and technology education programs at colleges and universities. ABET periodically reviews the contents of programs in accredited departments. Even if programs are not accredited, however, colleges and universities can still offer engineering programs and grant degrees.

4. *American Society for Engineering Education* (ASEE)—responsible for assisting in the exchange of the latest educational information and developing ties between universities and industry.

Again, due to the wide variety of organizations to which engineers belong, the enforcement powers of engineering as a profession have been somewhat limited.

3.6.2   A Short History of Codes of Ethics

Along with informal social sanctions, codes of ethics are the primary means through which professions control the actions of their members. In some

---

43. For more on the nature and development of the medical profession in the United States, see Ham and Alberti (2002).
44. Again, for more on the NSPE, readers are referred to its website at http://www.nspe.org.

professions, these codes have the force of law, since the state has designated them as such. Nurses in the United States, for example, can be prosecuted and punished for violating their professional codes of ethics, no longer able to practice nursing. Due to the fragmented nature of the profession, however, this is not the case in engineering. Nevertheless, engineering codes of ethics are visible symbols of the profession's commitment to the public good. To better understand the ideals underlying engineering codes of ethics, it is necessary to turn to ancient times.

In ancient Greece, Hippocrates developed one of the most famous codes for medicine. Its well-known introduction begins, "first, do no harm." Even before that, however, a code existed with implications for engineers: in 1758 BC, the Babylonian king established laws for civil engineers in the Code of Hammurabi, based on an "eye-for-an-eye" philosophy: "if a builder has built a house for a man and has not made his work sound and the house which he has built has fallen down and so caused the death of the householder, that builder shall be put to death."

In more recent times, 1847, the American Medical Association institution-alized a code of ethics, to which to the development of engineering codes can be traced back. In the early 20th century, the first US engineering codes of ethics were based on the medical code. The development of these codes was a rather obvious attempt to share in the prestige associated with the medical profession and, for this reason, they reflected the perspective of the medical code almost in full. The greatest consequence of this perspective is that engineers were conceived as independent practitioners in serving clients, rather than as employees working in companies. Additionally, these codes stressed the importance of fraternal relations between engineers. It was almost half a century before the ideal of public service came to prominence in engineering codes of ethics.[45]

A variety of engineering codes now exist, although most are similar to the ASME code of ethics (ASME, 2012a). For this reason, the rest of the discussion here focuses on the ASME code of ethics as an example. In examining this code, its historical nature should be kept in mind; it results from a variety of compromises. Additionally, as with other engineering codes of ethics, the ASME code contains propositions that potentially conflict with each other:

- Identify an occupation other than engineering that might be considered a profession but is currently not. Explain why and how it could be regarded as a profession.
- Do you think one professional organization responsible for overseeing the whole of engineering—like the American Medical Association for medicine—is feasible? Why or why not?

---

45. For more on the nature and history of codes of engineering ethics, see Luegenbiehl (1983) and Davis (2001).

3.6.3 The Contents of Codes of Ethics: ASME as an Example

The ASME Code of Ethics is comprised by ideals, principles, rules, guidelines, and rights. It has been used in a variety of ways: for the sake of professionalization, protection of group interests, teaching etiquette, inspiration and education, enforcement, and public relations. The focus here will be on the ethical content of the ASME code. The code consists of three main parts: (1) the Fundamental Principles, (2) the Fundamental Canons, and the (3) Criteria for Interpretation of the Canons (ASME, 2012b).

The Fundamental Principles section describes ideals toward which engineers should aspire. These principles are broad in scope and represent to the public the engineering profession. The following is the Fundamental Principles section of the ASME code in full:

"Engineers uphold and advance the integrity, honor, and dignity of the engineering profession by

I. using their knowledge and skill for the enhancement of human welfare;
II. being honest and impartial, and serving with fidelity their clients (including their employers) and the public; and
III. striving to increase the competence and prestige of the engineering profession."[46]

Compared with the Fundamental Principles, the Fundamental Canons are more rule-like in their assertions. There are 10 canons in total, and they address the following: the fundamental responsibility of engineers to maintain public safety, the environment, requirements of competence, honesty, loyalty, and fairness, as well as duty to support the profession and professionalism. The following is the Fundamental Canons section of the ASME code in full:

1. "Engineers shall hold paramount the safety, health, and welfare of the public in the performance of their professional duties.
2. Engineers shall perform services only in the areas of their competence; they shall build their professional reputation on the merit of their services and shall not compete unfairly with others.
3. Engineers shall continue their professional development throughout their careers and shall provide opportunities for the professional and ethical development of those engineers under their supervision.
4. Engineers shall act in professional matters for each employer or client as faithful agents or trustees and shall avoid conflicts of interest or the appearance of conflicts of interest.
5. Engineers shall respect the proprietary information and intellectual property rights of others, including charitable organizations and professional societies in the engineering field.

---

46. The ASME Code of Ethics can be found in its entirety at the ASME's website: https://www.asme.org/getmedia/9EB36017-FA98-477E-8A73-77B04B36D410/P157_Ethics.aspx.

6. Engineers shall associate only with reputable persons or organizations.
7. Engineers shall issue public statements only in an objective and truthful manner and shall avoid any conduct that brings discredit upon the profession.
8. Engineers shall consider environmental impact and sustainable development in the performance of their professional duties.
9. Engineers shall not seek ethical sanction against another engineer unless there is good reason to do so under the relevant codes, policies, and procedures governing that engineer's ethical conduct.
10. Engineers who are members of the society shall endeavor to abide by the constitution, bylaws, and policies of the society, and they shall disclose knowledge of any matter involving another member's alleged violation of this code of ethics or the society's conflicts of interest policy in a prompt, complete, and truthful manner to the chair of the ethics committee."[47]

Related to the ASME's Code of Ethics is its criteria for interpretation of the Canons. These give more detailed interpretations of the canons, to provide engineers with guidance in how to interpret the canons and to provide the profession with specific, enforceable entries. Unlike either the principles or the canons, the guidelines are rather long and go on for a number of pages. Thus, the following is merely a sample from the ASME's criteria for interpretation of the Canons, included to give readers a sense of their specificity:

"1.c.(2)Engineers shall conduct reviews of the safety and reliability of the designs, products, or systems for which they are responsible before indicating preliminary acceptance and before giving their approval to the plans for the design...
1.d.If engineers have knowledge of or reason to believe that another person or firm may be in violation of any of the provisions of these Canons, they shall present such information to the proper authority in writing and shall cooperate with the proper authority in furnishing such further information or assistance as may be required...
2.a.Engineers shall undertake responsible charge of engineering assignments only when qualified by education and/or experience in the specific technical field of engineering involved...
4.e.Engineers shall neither solicit nor accept gratuities, directly or indirectly, from contractors, their agents, or other parties dealing with their clients or employers in connection with work for which they are responsible. Where official public policy or employers' policies tolerate acceptance of modest gratuities or gifts, engineers shall avoid a conflict of interest by complying with appropriate policies and shall avoid the appearance of a conflict of interest..."

---

47. Again, the ASME Code of Ethics can be found in its entirety at the ASME's website: https://www.asme.org/getmedia/9EB36017-FA98-477E-8A73-77B04B36D410/P157_Ethics.aspx.

4.j.Engineers shall treat information coming to them in the course of their assignments as confidential and shall not use such information as a means of making personal profit if such action is adverse to the interests of their clients, their employers, or the public...

4.m.Engineers shall admit their own errors when proven wrong and refrain from distorting or altering the facts to justify their mistakes or decisions...

7.d.Engineers shall issue no statements, criticisms, or arguments on engineering matters that are inspired or paid for by any interested party, unless they preface their comments by identifying themselves, by disclosing the identities of the party or parties on whose behalf they are speaking, and by revealing the existence of any financial interest they may have in matters under discussion.

7.e.Engineers shall be truthful in explaining their work and merit and shall avoid any act tending to promote their own interest at the expense of the integrity and honor of the profession or another individual...

8.a.Engineers shall concern themselves with the impact of their plans and designs on the environment. When the impact is a clear threat to health or safety of the public, then the guidelines for this Canon revert to those of Canon 1...."[48]

- What do you think legitimizes codes of ethics, causing members of professions to adhere to them?
- List and explain both advantages and disadvantages engineers active in professional organizations might face versus engineers not active in such organizations? Explain your answers.

CASE STUDY TWO—GLOBAL PROFESSIONALISM? CHINA AND INDIA

In this chapter, establishing the relationship between professionalism and ethics, discussions have mainly focused on the nature and examples of professionalism in the United States. However, some have argued that grounding ethics in professionalism is inappropriate, since the notion and nature of professionalism is a largely US phenomenon and, therefore, inappropriate as a basis for applied ethics in cross-cultural and international contexts (Luegenbiehl & Fudano, 2005; Didier, 2010).

Although engineers should be sensitive to the environments in which they find themselves, professionalism should not be construed as a uniquely US phenomenon. In fact, the ambiguous nature of engineering as a profession and the consequences that follow from this ambiguity are a relatively universal phenomenon. To support this claim, the following considers the state and nature of professionalism in two of the world's most populous and quickly developing countries, China and India. China and India now graduate and employ more

---

48. The complete criteria for interpretation of the Canons can be found at the ASME's website: https://www.asme.org/getmedia/6e30b7a8-1be2-452a-83ec-9330d06175c8/Criteria_Fundamental_Canons.aspx.

engineers than any other countries. Better understanding the natures, while examining examples, of engineering professionalism in these countries helps to support the centrality of professionalism to a global account of engineering ethics.

Engineering Professionalism in China[49]

As in the United States, in China the terms "engineer" (工程师) and "engineering" (工科/理工) are somewhat ambiguous: no clear definitions or definite criteria exist regarding what it would mean to be an engineer or count as engineering, although the terms are tied to "science and technology" (科技). In China, the term "engineer" denotes a job title more than a profession. For example, like the computer industry in general and the field of software development specifically, one who holds a degree in computer science might be referred to as a "software engineer," although this would be used as a job title rather than a professional moniker. Hence, as in the United States, in China, the professional nature of engineering is fragmentary, originating from and evident in university education to professional organizations and government-industry initiatives.

Considerable variation exists in the organization of Chinese universities, affecting the institutional standing of engineering. In Peking University, for instance, departments fall under broader categories, such as science, engineering and technology, and art and humanity. In Tsinghua University, by contrast, departments fall under specific subjects, such as math, physics, chemistry, and mechanics. Hence, although engineering is related to science and technology, clear boundaries between them do not exist. The ambiguous status of engineering as a profession within China is also evident in the nature of its professional organizations.

Founded in 1994 and under the direct administration of the State Council, the Chinese Academy of Engineering (henceforth the CAE) is the highest academic institution in the field of science and technology. However, the CAE does not directly oversee either research institutes or state labs and, thus, does not carry out research, unlike, for example, the Chinese Academy of Science. Rather, the main role of its members is to provide expertise guidance regarding potential engineering programs carried out by the State Council. For this reason, membership in the CAE is highly selective and honorary in nature. Traditionally, CAE members have been based in research universities and institutes although, as of late, the ratio of members from the private sector and business has increased. Membership is based on approval by the whole of the CAE, and the nomination of new members is proposed by current members of the CAE or China Association for Science and Technology (hereafter the CAST).[50]

---

49. This refers to the People's Republic of China (mainland China) rather than the Republic of China (Taiwan).
50. Information regarding the CAE can be found at its website, including a list of members and their educational background and work experience: http://www.cae.cn.

Unlike the CAE, which is an elite advisory organization consisting in honorary membership, the CAST is primarily a social organization with millions of members—although most individuals holding executive positions have government backgrounds. The CAST was established in 1958, resulting from the combination of two previous organizations focused on exchanging knowledge between scholars and distributing knowledge to the public, respectively. Based on its origins, today the CAST is responsible for providing educational assistance programs, transmitting knowledge to the public, and facilitating the exchange of knowledge within industries. For example, in collaboration with the Ministry of Education, the CAST organizes educational programs and competitions for students—such as an award program for future scientists—activities for the exchange of academic information for scientists and engineers, and science events for the general public. These programs and competitions aim at integrating academic education and research with industry. Hence, similar to technical engineering organizations in the United States, the CAST has a variety of suborganizations associated with different specialties, such as the Chinese Mechanical Engineering Society (CMES) and Chemical Industry and Engineering Society of China (CIESC). Unlike technical engineering organizations in the United States, however, the CMES, CIES, and other suborganizations within the CAST are not responsible for developing technical standards for industry.[51] Rather, these are the responsibilities of the Ministry of Industry and Information Technology (henceforth the MIIT).

The MIIT is responsible for developing industrial production standards: it sets all official industrial and technical standards in China. Having been founded in 2008—as a result of departmentalization reforms in the State Council—the MIIT is a relatively new institution (Xu, 2008). Since the Chinese government takes an active role in markets, the MIIT is also responsible for determining the technical directions toward which industries should move. Information technology is particularly important, such that the MIIT controls communication networks, tasked with safeguarding information. Hence, within the field of information technologies, the MIIT is also responsible for foreign cooperation and exchange. In 2016, for example, it organized a group to attend and present at the Consumer Electronics Show in Las Vegas, the United States, both to present Chinese research and learn about that of other countries[52]:

- What advantages and disadvantages might result from fluid definitions and understandings of engineering?
- What advantages and disadvantages might result from the MIIT—rather than organizations associated with engineering specialties—determining industrial standards?

---

51. Information about the CAST, including organized events, can be found at its website: http://www.cast.org.cn.
52. Information concerning the MIIT, including that regarding its foundation, functions, and members can be found online: http://www.miit.gov.cn. For more on the nature of engineering practice and ethics in China, see Zhu (2010).

Engineering Professionalism in India

Given its colonial past under British rule—and the subsequent development of institutions based on this past—the organization of engineering in India is closer to its North American and European counterparts than that of China. Examining this development sheds light on the nature of engineering in India today.

Professionally, the Institution of Engineers India (henceforth the IEI) is the oldest engineering organization in India, established in Calcutta (present-day Kolkata) on Sep. 13, 1920. It gained prominence when granted a royal charter by King George V in 1935. Educationally, the first Indian Institute of Technology (IITs) was founded in Kharagpur (West Bengal) by parliament in 1956, although engineering did not become popular as a field of education and occupation until the 1990s (Karnik, 2015).

To a large extent, this increase in popularity can be explained with reference to changes in India's economy: the expansion of foreign trade in the private sector, resulting in greater national importance attached to software services, technology outsourcing, building construction, etc. This emphasis has resulted, in turn, in greater demand for specialists and experts in the field of engineering (Karnik, 2015). Meeting this demand has helped to create a better, more fluid Indian economy, but has also required greater oversight to ensure the quality of engineering education and engineering as a profession.

Within education at the IITs, the IEI has recognized programs in agricultural, electrical, mechanical, and metallurgical engineering, as well as naval architecture and marine engineering ("Historical Events"), and the National Institute of Technology (NITs) was founded to administer regional colleges by the Indian government in 2002. In Aug. 2014, the Indian Ministry of Human Resources and Development (MHRD) brought together directors of IITs and NITs, establishing a committee to set up a national ranking framework for engineering and business institutions.[53] In terms of professionalism, the IEI is a signatory of the International Professional Engineers Agreement (IPEA), with bilateral agreements between a variety of national and international engineering institutions.

To become a professional engineer (PE) or an international professional engineer (IntPE) in India, an applicant must hold a Bachelor of Engineering (BE), Bachelor of Technology (BTech), or an equivalent degree—for example, a Bachelor of Science (BSc)—from an institution recognized by the Indian government or a statutory authority. Additionally, an individual should have at least 7 years experience in his or her field, of which two should be in positions of responsibility devoted to significant engineering activity ("Professional Engineers"). The Board for Certification, composed of representatives from national professional institutions, reviews and decides on applications.[54]

---

53. For more information, see https://www.nirfindia.org/Home.
54. This process is similar in nature to that of becoming a PE in the United States, as established by the NSPE, where, in general, an applicant should (1) hold an engineering degree from an ABET accredited program, (2) pass the Fundamentals of Engineering exam, (3) acquire 4 years professional experience, and (4) pass the Principles and Practice of Engineering exam in the licensing state ("How to Get Licensed").

Although the profession of engineering in India is similar to that in many Western countries, as a profession, engineering exists in India among other features of social organization, one of which is the caste system.

With premodern roots, the Indian caste system is a form of institutionalized social stratification based on birth, where those occupying higher castes have traditionally had greater opportunities—in terms of work, marriage, etc.—than those occupying lower castes. The Congress party, formed approximately 200 years ago, has run on ideals of equality and progress, coming to power and forming coalition governments by promising greater opportunities to those marginalized within Indian society. This has lead to the creation of quotas for Scheduled Castes (SC) and Scheduled Tribes (ST) for spots in institutions of national education and jobs in government, influencing the profession of engineering throughout the country. As with programs of affirmative action in the United States, many in India worry that this quota system results in less-qualified candidates receiving positions based on their castes.[55] Additionally, many multinational corporations have set up programs to address inequalities associated with the caste system in India. The software giant Infosys, for instance, has set up programs to train individuals from lower castes, and the company often hires candidates from these programs.

3.7 SUMMARY

As the case at the beginning of this chapter demonstrates, versus mere occupations and occupational workers, society has higher expectations of professions, professionals, and professional organizations, holding them to higher standards. In part, this results from the fact that society provides professionals with relatively high prestige, pay, authority, and autonomy. This would be part of a contract-model understanding of professions, where society gives these to professionals in exchange for professionals providing society with indispensable services and insuring the quality of these services. Professions ensure this quality through the control of its members, in the form of the licensing, accreditation, and codes of ethics. Although engineering has gone a long way in establishing itself as a profession, its status is still ambiguous: neither has engineering a monopoly on the service of engineering nor does it solely determine who can or cannot be considered an engineer. This organization of engineering as a profession is not limited to the United States. Engineering exists in the form of a profession—with complex relations between individual engineers, engineering organizations, and society—in two of the world's most populous, quickly developing countries, China and India.

55. For more on India's caste system, see Business and Caste in India: With Reservations (2007).

REVIEW QUESTIONS

1. List and explain the characteristics of Hydrolevel's boiler cutoff valves that allegedly made them unsafe.
2. In which companies did Hardin hold positions and how did he take advantage of holding these positions simultaneously?
3. Explain how the Hippocratic Oath and Code of Hammurabi have shaped modern-day understandings of ethics for engineers.
4. List the parties involved in the contract model of professionalism and explain the relations between these parties.
5. What roles do codes of ethics play in the relationship between individual professionals, professional groups, and society?
6. Describe how engineering has developed as a profession since the mid-19th century.
7. Name the three main components of the ASME Code of Ethics, explaining how they function.
8. Explain similarities and differences between the development of engineering as a profession in China and India and potential reasons for these differences.

REFERENCES

ASME. (2012a). ASME code of ethics. https://www.asme.org/getmedia/9EB36017-FA98-477E-8A73-77B04B36D410/P157_Ethics.aspx.

ASME. (2012b). ASME criteria for interpretation of the canons. https://www.asme.org/getmedia/6e30b7a8-1be2-452a-83ec-9330d06175c8/Criteria_Fundamental_Canons.aspx.

Bucher, R., & Stelling, J. (1969). Characteristics of professional organizations. *Journal of Health and Social Behavior, 10*, 3–15.

Bucher, R., & Strauss, A. (1961). Professions in process. *American Journal of Sociology, 66*, 325–334.

Business and Caste in India: With Reservations. (2007). *The Economist*, October 4. http://www.economist.com/node/9909319

Davis, M. (2001). Three myths about codes of engineering ethics. *IEEE Technology and Society Magazine*, Fall.

Davis, M., Jones, H., & Wells, P. (1986). *Conflicts of interest in engineering*. Dubuque, IA: Kendall/Hunt Publishing Company.

Department of Philosophy and Department of Mechanical Engineering. (n.d.). Texas A&M University. American Society of Mechanical Engineering (ASME) vs. Hydrolevel Corp. http://ethics.tamu.edu/Portals/3/Case%20Studies/ASMEVersusHydrolevelCorp.pdf

Didier, C. (2010). Professional ethics without a profession: A French view of engineering ethics. In I. Van de Poel & D. Goldberg (Eds.), *Philosophy and engineering: An emerging agenda* (pp. 161–174). Dordrecht: Springer.

Fitzgerald, G. (2005). Engineering perceptions—Raising the profile of the "invisible profession". *The Engineers Journal, 59*(8).

Galanter, M. (1993). Predators and parasites: Lawyer-bashing and civil justice. *Georgia Law Review, 28*, 633.

Galanter, M. (1997). Faces of mistrust: The image of lawyers in public opinion, jokes, and political discourse. *The University of Cincinnati Law Review, 66*, 805.

Ham, C., & Alberti, K. (2002). The medical profession, the public, and the government. *BMJ, 324,* 838.

Harris, C., Pritchard, M., & Rabins, M. (2009). *Engineering ethics: Concepts and cases* (4th ed.). Belmont: Wadsworth.

Hensler, D. (2003). Our courts, ourselves: How the alternative dispute resolution movement is reshaping our legal system. *Pennsylvania State Law Review, 108,* 165.

Historical Events of the Institution. (n.d.). The Institution of Engineers (India). https://www.ieindia. org/PDF_IMAGES/royalchar/HistoryIEI.pdf.

How To Get Licensed. (n.d.). NSPE. http://www.nspe.org/resources/licensure/how-get-licensed.

Karnik, M. (2015). India's engineering boom, 1995 to the present. *Quartz,* 22 April. http:// qz.com/388557/indias-engineering-boom-1995-ongoing/.

Larson, M., & Larson, M. S. 1979. *The rise of professionalism: A sociological analysis.* Los Angeles, CA: University of California Press.

Layton, E., Jr. 1986. *The revolt of the engineers: Social responsibility and the American engineering profession.* Baltimore, MA: Johns Hopkins University Press.

Luegenbiehl, H. (1983). Codes of ethics and the moral education of engineers. *Business & Professional Ethics Journal, 2*(4), 41–61.

Luegenbiehl, H., & Fudano, J. (2005). Japanese perspectives. In C. Mitcham (Ed.), *Encyclopedia of science, technology, and ethics.* Detroi, MI: Macmillan Reference.

Professional Engineers & International Professional Engineers Certification. (n.d.). The Institution of Engineers (India). https://www.ieindia.org/PEInfo.aspx.

Re, E. (1994). The causes of posular dissatisfaction with the legal profession. *St. John's Law Review, 68,* 85.

Xu, P. (2008). Guo Wu Yuan Ji Gou Gai Ge Fang An (The Departmentalization Reforming Project of the State Council). *Xin Hua News.* 15 March. http://news.xinhuanet.com/misc/2008-03/15/content_7794932.htm.

Zhu, Q. (2010). Engineering ethics studies in China: Dialogue between traditionalism and modernism. *Engineering Studies, 2*(2), 85–107.

FURTHER READING

Xu, X. (2000). *Chinese professionals and the republican state: The rise of professional associations in Shanghai 1912–1937.* New York: Cambridge University Press.

Chapter 4

Basic Ethical Principles for Global Engineering

Chapter Objectives

Having read this chapter, completed the included exercises, and answered the associated questions, readers should be able to

- give examples of why the need exists for broad but commonly agreed upon principles of global engineering ethics, with reference to the case of Ford and Firestone/Bridgestone;
- explain problems associated with pregiven engineering ethics codes, how the approach here is different, and why/how the safety of human life plays a central role in engineering ethics;
- list the first 6 basic ethical principles for global engineering and justify their derivation based on the primacy of safety, as well as identify instances in which they are relevant in the cases of Ford and Firestone/Bridgestone and Development and its Broader Contexts.

CASE STUDY ONE—FORD AND FIRESTONE/ BRIDGESTONE: RESPONSES TO TECHNOLOGICAL FAILURES

On Jan. 11, 2001, Yoichiro Kaizaki announced his resignation as president, chairman of the board, and chief executive officer of the Bridgestone Corporation ("Bridgestone Names," 2001), thus bringing about a Japanese point of resolution to the ongoing Ford-Firestone tire crisis. His resignation came after a scandal involving the separation of tread on Firestone tires, allegedly responsible for over 200 deaths and 3000 serious injuries. Kaizaki's resignation was somewhat atypical: "As far as I'm concerned, this is not taking responsibility for the recall," he said, referring to it as "a bit of trouble" (Dvorak & Williams, 2001). His resignation reflects traditionally typical and atypical Japanese responses to business crises: Kaizaki ultimately resigned from his position in an apparent attempt to atone for the actions of Bridgestone, without actually acknowledging responsibility.

This incident and the responses by both Ford and Firestone/Bridgestone highlight the need for broad but uniform ethical principles in cross-cultural and

Global Engineering Ethics. http://dx.doi.org/10.1016/B978-0-12-811218-2.00004-7
53

international engineering and business contexts. Ford and Firestone are both old, well-known US companies, although Firestone is a wholly owned subsidiary of the Japanese Bridgestone Corporation, and its top leadership at the time of this crisis consisted mainly of Japanese executives.

Those working with technology should recognize that their actions increasingly take place in global environments with corresponding ramifications, although developers and producers of technology often have narrower vision, perceiving such issues through the lenses of localized ethical and value systems alone. The Firestone-Ford case provides an excellent example of how restricted national and cultural viewpoints can influence both the implementation of technology dispersal strategies and the responses to the failure of technology.

General Case Background: Responses to Public Pressure

This was not the first time Firestone had been involved in a major recall: in 1978, Firestone recalled between 13 and 14 million Firestone 500 tires, as a result of blowouts and tread separation. This cost the company $200 million (ElBoghdady, 2000). Fitted to mostly Chevrolets, the recalled tires were linked to 41 deaths, although Firestone fought the recall for months (Aeppel, 2001). As a result of the costs associated with the recall and negative publicity, the company was weakened considerably, to the extent that in 1988 it was sold to the Bridgestone Corporation for $2.6 billion ("Analysis," 2000).

In 1990, Ford began producing its extremely successful "Explorer" sport-utility vehicle (SUV) and, in 1991, fitting the Explorer with Firestone Wilderness tires. That same year, Kaizaki was appointed head of the US Bridgestone/Firestone subsidiary, tasked with turning around the financials of the company (Kunii & Foust, 2000). In 1992, the first lawsuit related to Firestone tires tread separation was filed in the United States (Eisenberg, 2000). Having improved the financials of the US Bridgestone/Firestone subsidiary, Kaizaki was rewarded in 1993, being named CEO of Bridgestone, the parent company in Japan (Eisenberg, 2000). Masatoshi Ono was appointed as Kaizaki's replacement at Firestone. Ono had been with Bridgestone since 1959 and was sent to the United States as Kaizaki's second-in-command in 1991 (Zaun & Shirouzu, 2000).

From 1994 to 1996, major strikes occurred at a Firestone plant in Decatur, Illinois, during which time the company fought the unions as part of Firestone's cost-cutting efforts (Merrick, 2000). Subsequent analyses identified production problems at this plant during this period of time, specifically, as one of the sources of tire failures, although the degree of causation has been in dispute. In 1996, Firestone conducted extensive tests on the later recalled tires, making changes to the design of the tire to improve stability. The company denied that these changes were connected with the testing (Fogarty & Eldridge, 2000).

In 1998, Sam Boyden, an employee of the State Farm Insurance Company, notified Firestone and the National Highway Traffic Safety Administration (NHTSA) that he was encountering an abnormal claims pattern on certain Firestone tires (Spurgeon, 2001). In that same year, Ford recognized a pattern of tire failures in Venezuela. In 1999, Ford began a tire replacement program in Saudi Arabia—which they termed a "customer notification enhancement action"—without notifying either the public or the NHTSA (Simison, Shirouzu, & Aeppel, 2000). Subsequently, the replacement program widened to include 16 foreign countries (Pickler, 2000b). The replacement of tires in Venezuela began in May 2000 (Simison, Shirouzu, & Aeppel, 2000).

On Feb. 7, 2000, a series of investigative reports began being aired by KHOU, a television station in Houston, first bringing the issue of tread separation on Firestone tires to the attentions of the US public. On Feb. 10, Christine Karbowiak, vice president for public affairs at Firestone, responded to the broadcasts: "This series has unmistakably delivered the false messages that Radial ATX tires are dangerous, that they threaten the safety of anyone using them, and that they should be removed from every vehicle on which they are installed. Each of these messages is simply untrue. This is a good product and Firestone proudly stands behind it." She further wrote that the cases reported on were "clearly caused by external factors, such as punctures," that the sources used for the investigative reports were employees who left Firestone "disgruntled and unhappy," and that the television station "would better serve" its "viewers in the Houston area if" it "would point out to them proper tire maintenance procedures" ("Firestone," 2000). This became a refrain of Firestone's after the recall in the United States was announced.

On Apr. 30, the *Chicago Sun-Times* published a report on tire blowouts, although it did not address Firestone tires specifically (Skertic, 2000). This report is sometimes credited with spurring the NHTSA to action, which sent a defect investigation letter to Firestone on May 8, in which the NHTSA indicated that it had received 90 complaints involving Firestone tires, including 33 crashes, 27 injuries, and 4 deaths ("Tire Failures," n.d.). At approximately that time, Ford began its own investigation into the tire failures, and on Jul. 28, it received warranty claims data on Firestone tires and began analyzing the data (Simison, Shirouzu, Aeppel, & Zaun, 2000). On Aug. 3, *USA Today* reported that "Ford Motor... would consider dropping Bridgestone/Firestone as a tire supplier if consumers balk at buying vehicles equipped with Firestone tires involved in a government safety investigation." The report also raised the number of claims being investigated to 193 crashes and 21 deaths (Healey & Nathan, 2000a, 2000b). The next day, Sears, a US department store chain, announced that it would stop selling certain Firestone tires ("Sears," 2000).

On Aug. 8, after a meeting between the NHTSA, Ford, and Firestone, the companies decided on a recall, to be initiated Aug. 9. All P235/75R15 Firestone Radial ATX and ATXII tires were recalled, as well as all P235/75R15 Firestone Wilderness AT tires produced at the Decatur plant (Pickler, 2000a). 14.4 million of these tires had been produced in total, and Firestone estimated that approximately 6.5 million remained on the road. In announcing the voluntary recall, Gary Crigger, the executive vice president of Firestone, said, "first let me say that, at Bridgestone/Firestone, nothing is more important to us than the safety of our customers." Firestone indicated that the recall would take place in phases, over a 1-year period, beginning with the warmest parts of the country ("Statement by Christine," 2000).

Throughout the country, Explorer owners immediately began demanding replacement tires, which Firestone was unable to accommodate (Schneider, 2000). By Aug. 16 and from the time of the recall, the NHTSA had raised the number of incidents linked to the tires from 270 complaints, 80 injuries, and 46 deaths to 750 complaints, 100 injuries, and 62 deaths (Truby & ElBoghdady, 2000). By Feb. 2001, those numbers had risen to more than 6000 complaints, more than 700 injuries, and 174 deaths (Pickler, 2001).

As a result of panic among consumers and actions taken by advocacy groups, Firestone quickly changed the terms of its initial recall: Firestone included the possibility of consumers installing alternative tire brands (Grimaldi, 2001). By Aug. 30, 1 million tires had been replaced ("Fourth," 2000). By the middle of Sep., Firestone announced that it expected to finish the recall by the end of Nov. 2000 (Dreazen, 2000).

The Conflict Between Firestone and Ford: She Said, He Said

There are numerous ethical issues one could explore in relation to this case, including, for example, quality control, customer relations, and government involvement. To highlight questions of cross-national and cross-cultural ethics and values, however, and the corresponding need for ethical principles in global engineering and business contexts, the following focuses on the relationship between Ford and Bridgestone/Firestone specifically. That relationship has a long history, going back to purchases by Henry Ford from Harvey Firestone in 1906 (Kiley & Healey, 2000). At the time of the recall, Ford was Firestone's largest customer, although, beneath the surface, strategy disagreements were already apparent.

On Aug. 4, 2000, Jacques Nasser, the president of Ford, said in a statement that "we're working extremely closely with the US government and Firestone because we want to get to the bottom of this (tire issue) as quickly as we can" ("Jac," 2000). Ford and Firestone had in fact signed a joint agreement, according to which they cooperated in defense against previous rollover suits, which generally pointed to "worn tires or driver error" as the source of accidents (Geyelin, 2000). Later, however, it was discovered serious disagreements existed between the two companies regarding proposed recall actions in Saudi Arabia and Venezuela. In both cases, Ford eventually unilaterally replaced Firestone tires on Explorer vehicles ("Statement by Gary," 2000; Simison, 2000). According to Ford, in each instance, it had asked Firestone to conduct tests on their tires and those sold in the Southwest United States and was told by Firestone that "there was no defect" (Ford Motor Company, 2000). Firestone confirmed this in a statement by Crigger: "Nothing we learned...led us to believe we had a defect with these tires" (Power & Simison, 2000).

Soon after initiation of the recall, disagreements between Ford and Firestone became public. On Aug. 11, Ford authorized dealers to replace Firestone tires with other brands ("Ford Authorizes," 2000). Signaling his intention to blame Firestone, Nasser said there is "absolutely no data or incidents of trends of incidents" showing design problems in the Explorer (Kiley & Healey, 2000). As public anger over the slow pace of tire replacement grew, Nasser stated even more forcefully: "This is a tire issue, not a vehicle issue" (Pickler, 2000a). In the meantime, Firestone spokespersons said "the biggest task in front of us is preservation of our business with Ford" and keeping "the Firestone brand alive" (Shirouzu & Aeppel, 2000). By Sep. 6, when the first Congressional hearing on the issue occurred, the division between the two companies was clear.

Nasser and Ono sat at separate witness tables (Shirouzu & Power, 2000). Nasser stated that Ford "virtually pried the claims data (for the recalled tires) from Firestone's hands" and that, once they received the data, Ford engineers "discovered conclusive evidence that the tires were defective" (Ford Motor Company, 2000). On Sep. 11, in his first public statement, Bridgestone's Kaizaki responded indirectly: "It's difficult to specify the causes of the accidents.... But I wonder why most of the accidents happened to Explorer models with the tires?" ("Bridgestone Will," 2000). In another Japanese interview, he added, "we faithfully kept our side of the deal, but it was all for naught. Holding our tongues was a mistake" ("Bridgestone President," 2000). Ford responded by announcing that alternative tires would be made available for 2002 model Explorers. In this announcement, Nasser made it clear that Ford had suffered as a result of the ongoing recall and the part played by Bridgestone: "This has been an extremely difficult and disappointing time in our relationship. We're going to evaluate it a day at a time" (Eisenberg, 2000).

As time passed, however, a noticeable shift in public opinion occurred, as it became clear that Ford played an active role in the original testing of the tires and had misstated information to Congress regarding the testing procedures (Kiley & Healey, 2000; Dobbyn, 2000). Hence, as the recall was being completed and the companies again began to focus on the eventual legal liability resulting from this case, their statements became more conciliatory (Aeppel, Power, & White, 2000). Nasser expressed himself as "confident that (the NHTSA), Firestone, and Ford will get to a common understanding and conclusion." He added "I'd classify the relationship as a good businesslike relationship." Firestone, in turn, admitted some fault: "We're looking at the way the vehicle and the tire may operate as a system" (Zaun, White, & Aeppel, 2000).

The Nature of the Disagreements: Tire Pressure and Leadership Styles

As indicated above, significant tensions existed between Ford and Firestone. What were the sources of these tensions and, more importantly, how do they indicate differences in and confusions regarding ethics and values?

Tire Pressure. A major, public source of conflict was the recommended pressure for tires on the Explorer. At the time of the recall, Firestone stated the following in a press release: "We urge all vehicle owners using ATX and Wilderness tires to keep your tires inflated at the pressure recommended by the vehicle manufacturer." It added, however, that while Ford recommended a pressure of 26–30 psi (pounds per square inch), Firestone recommended 30 psi. The press release further emphasized that most of the incidents Firestone had reviewed were the result of underinflation (Pickler, 2000a). In a public statement, Firestone's Ono added, "there is no doubt it's better to set a tire-inflation rate higher under extremely hot weather, but we do not believe Ford's recommendation of 26 psi for our tires was a mistake" (Shirouzu & Aeppel, 2000).

Soon it became apparent that inflation was a crucial part of the dispute between the two companies, and that seemingly contradictory information added to customers' confusion and anger. On Sep. 20, Firestone sent a letter to Ford, recommending a tire pressure of 30 psi (Simison, Shirouzu, Aeppel, & Zaun, 2000). Two days later, Ford raised its recommendation to 30 psi, citing confusion among its customers as the reason for the change (Hyde, 2000). At the same time, however, Ford noted that both companies had supported the 26 psi recommendation for over 10 years, and that an equivalent recommendation for Goodyear tires had not resulted in any problems (Aeppel, Power, & Geyelin, 2000). Later investigation revealed that Ford's motive for the initial recommendation was based on the need to increase the stability of the Explorer, and it had significantly reduced the margin of tire safety (Aeppel, 2000b).

Leadership. Another source of conflict was differences in approaches taken by the two companies' leadership in response to the crisis. Nasser's initial public statement was as follows: "You have my personal guarantee that all the resources of Ford Motor Co. are directed at resolving the situation. There are two things we never take lightly—your safety and your trust." He also announced that Ford would idle three plants producing Explorers, so tires from those plants could be diverted to the replacement effort, and that Ford would focus on cost sharing with Firestone at a later time ("Bridgestone/Firestone," 2000). In contrast, during a phone interview, Firestone's Ono said, "I am leading a life pretty much as usual" and that the recall was an "ordinary one when a problem like this occurs" (Simison, Shirouzu, Aeppel, & Zaun, 2000). At a Congressional hearing on Sep. 6, in his first public appearance during the recall, since an initial meeting with reporters, Ono issued an apology to the public and accepted "full and personal responsibility." He emphasized again, however, that problems resulted from a "lack of care for the tires. That would be my conclusion." On the other hand, Nasser attacked Firestone, blaming it for producing defective tires (Mayer & Swoboda, 2000).

During the initial phases of the crisis, both William Clay Ford Jr., chairman of Ford, and Kaizaki, chairman of Bridgestone, were conspicuously absent from the discussions. When asked about this later, William Ford said, "I think it would have been very confusing for the company to have two spokesmen out at this very critical time" (Simison, 2000). In contrast, Kaizaki's justification was that "an explanation by a Japanese person who does not know the situation well could have gotten on US consumers' nerves" ("Analysis," 2000). Press reports explained Firestone's approach as resulting from "weak global management" ("Analysis," 2000) and a lack of understanding of the need for public relations (Kunii & Foust, 2000). Another explanation was that "like most Japanese executives who find themselves under scrutiny, he (Kaizaki) is lying low" (Kunii & Foust, 2000).

As the crisis progressed, it became increasingly clear that Firestone's leadership was overwhelmed by a more than 5 million dollar advertising campaign mounted by Ford ("Bridgestone/Firestone," 2000). On Oct. 20, Bridgestone announced that John Lampe would replace Ono as president of Firestone. Lampe had testified before Congress in a spirited defense of the company, in contrast to what Bridgestone officials considered to be a weak performance by Ono, whose remarks had to be translated into English ("Bridgestone Will," 2000). In becoming the first US president of Firestone, Lampe promised full disclosure to the public and accepted responsibility for the problems that had arisen: "And we know that we can't blame anyone else for people losing trust in Firestone products—not our customers, not our business partners, not the media or Congress. The responsibility is ours. I want my first act as the new CEO of Bridgestone/Firestone to be an apology to those who have suffered personal losses or who have had problems involving our products" ("A Message," 2000). In his statement of resignation, Ono said his departure was unrelated to the recall, instead pointing to his age and health problems as the reasons (Miller, 2000b). Similarly, several months later when Kaizaki resigned as head of Bridgestone, he refused to connect his resignation to the recall: "As far as I'm concerned, this is not taking responsibility for the recall" (Dvorak & Williams, 2001).

Crisis Management: Differences in Resource Allocations

The difference in approaches taken by the leadership of the two companies was also evident in the ways they assembled teams to deal with the crisis. Ford announced that it had formed a team of 500 employees specifically to address the crisis, with another 4000–5000 employees involved, and had established 24 h hotlines to answer customer questions, and Nasser was participating in daily crisis meetings (Simison, 2000). Ford emphasized that Nasser was in complete control of operations. In contrast, Bridgestone explained that Ono was absent from public view because he was in contact with employees and Ford. Indications were that Ono had delegated responsibility for the crisis to US managers and that he had assembled a team of 5–6 US executives to handle the task (Shirouzu & Aeppel, 2000). Bridgestone said that, in Japan, a team of 20 had been assembled to deal with the crisis and that Kaizaki had made a secret trip to the United States to assess the situation (Zaun, Dvorak, Shirouzu, & Landers,

2000). While Nasser appeared as pushing employees to work harder to satisfy Ford customers (Simison, 2000), Kaizaki was quoted as saying, "rebuilding this company is my biggest responsibility" (Zaun, 2000). In the US press, comments were made that failure in leadership was "a classic Japanese problem" (Shirouzu & Aeppel, 2000) and "Japanese companies often respond too late to a crisis" ("Opinion," 2000).

The Ambiguous Firestone/Bridgestone Relationship

Leadership initiatives were complicated by the fact that Firestone's relationship with Bridgestone was not fully transparent. Bridgestone was apparently surprised by the recall, having several weeks earlier predicted good profits (Greenwald, 2000). The Japanese press seemed to support this interpretation: a Bridgestone executive was quoted as saying, "as the parent company, we should have established a system whereby such accident information is reported immediately. We are responsible for failing to supervise the subsidiary" ("Bridgestone Forecasts," 2000). Another Japanese press report stated the following: "Aware that Americans would not listen to a person who worked with one eye fixed on what the parent company back in Japan was thinking, Firestone was given free rein to lay off workers as it saw fit and to revamp quality control from the bottom up." The same report said "in their expansion overseas, Japanese companies tend to appoint local hires to executive positions and entrust operations to the people on the spot" ("Bridgestone Fiddles," 2000).

However, Ono indicated that he was in constant contact with leadership in Japan throughout the crisis (Simison, Shirouzu, Aeppel, & Zaun, 2000), and a report in the US press quoted a former Firestone executive as saying that, after executive meetings with the same number of Americans and Japanese present, the Japanese would hold a separate meeting at which the actual decisions were made, the final authority always resting with Kaizaki (Kunii & Foust, 2000). Interestingly, when Lampe was named to lead Firestone, four executives from Bridgestone were also appointed to a new executive team meant to lead with him (Aeppel, 2000a).

The apparently contradictory nature of messages being sent by Firestone and from Japan added to the confusion. During the Congressional hearings, Lampe said the company had produced "some bad tires." Later, Kaizaki disputed this claim in the Japanese press, which also reported that "a Bridgestone spokeswoman in Tokyo said the company doesn't believe Mr. Lampe made such a comment, which was widely reported in the US media" (Dvorak & Williams, 2000). In the Japanese press, Bridgestone's response was said to have "angered the American public" ("Bridgestone Fiddles," 2000). Later, when he explained the parent company's actions, Kaizaki placed the blame on Ford: "We had decided to keep in step with Ford in the public relations system. However, Ford later broke ranks and unilaterally tried to force all responsibility onto Bridgestone" ("Missteps," 2000).

One proposed explanation for the mixed reactions from Firestone and Bridgestone was the deep "physical and psychological gulf" between the parent and its subsidiary, that people in Japan "don't think of them as the same company," despite the fact that over half of Bridgestone's profits came from its Firestone operations (Zaun & Dvorak, 2000). Traditional Japanese shareholder relationships shield companies from takeovers, and, thus, perhaps Bridgestone did not feel threatened by the recall (Spindle, 2000).

External Relations: The Press and Public

Moreover, Bridgestone did not seem to feel the need to deal with the press and public in the same way as Ford. In the United States, Firestone consistently emphasized that the consumer was a likely source of problems. In his initial public statement, Ono said that consumers should check tire pressure each month, and "it would be better if they" checked "it once every two weeks" (Shirouzu & Aeppel, 2000). As part of the initial recall announcement, Firestone wanted to simply read a statement without taking questions. Ford insisted otherwise (Simison, Shirouzu, Aeppel, & Zaun, 2000). When asked about the lack of public presence by Bridgestone during the crisis, a spokesperson said that executives had no time to talk to the press: "Bridgestone sees nothing to be gained by having its executives talk to the public" (Zaun & Dvorak, 2000).

In response to continuing consumer complaints, Bridgestone announced it would begin airlifting replacement tires from Japan, but—due to a shortage of the appropriate molds—this promise went largely unfulfilled (Zaun & Woodruff, 2000). When the *Wall Street Journal* sent a reporter to production plants in Japan, he found that "the atmosphere" was "anything but intense," with no extra shifts and little overtime scheduled. A taxi driver he interviewed said "Bridgestone is Bridgestone and Firestone is Firestone. Most people here don't think there is a connection" (Zaun & Dvorak, 2000).

A lack of interest in Japan also resulted from the fact that few Explorers were imported into Japan and, thus, the recall there was limited to 6020 tires. In Japan, no lawsuits were filed (Schneider, 2000), and only one recall on original tires had ever occurred—in 1976, involving 2839 vehicles. A partial explanation for this might be the fact that the Japanese Ministry of International Trade and Industry assigns only two regulators to the tire industry. In contrast, the US NHTSA assigns 47 regulators to tires and other automotive parts. Further, Japanese regulators lack any real authority or technical training, having to force their investigations through companies' public relations departments (Dvorak & Zau, 2000). One proposed explanation for this is the lack of an active consumer movement in Japan (French, 2000). Or, as a US auto analyst in Tokyo put it, "the general perception is it's probably Ford's fault because Americans make bad cars" (Zaun & Dvorak, 2000).

Confusing statements made by the Japanese leadership further complicated public relations. For example, Kaizaki said, "we didn't recall the tires because we found a defect that caused the accidents. We decided to conscientiously recall the tires having put a top priority on consumer safety" ("Firestone Spreads," 2000). In a deposition, Ono said, "at present, we have not concluded whether or not there was a defect. However, we have to acknowledge there may have been safety related problems—there were safety related problems" (Miller, 2000a). Firestone only considered a tire defective once a specific flaw had been found, and as long as it was conducting its internal investigation, no specific root cause of the tread separation had been identified. As the press reported, however, "the difference is technical, but muddies an already murky situation" (Aeppel, 2000b).

Ford took a different approach in its public relations campaign, publicizing its efforts to counter Firestone's limited actions. Ford emphasized customer relations, promising to release all relevant documents and cease production of the Explorer (Eisenberg, 2000). During the Congressional hearing, Nasser said, "originally, Firestone wanted to prioritize shipments of tires to certain states. During this meeting, we said 'that isn't customer-friendly. That just isn't right'" (Simison, 2000). As a result of public statements, Ford thus appeared the winner, despite the fact that Nasser had initially refused to attend the Congressional hearings (Spurgeon, 2000). Bridgestone seemed to ignore the human side of the equation: its public statements stressed the low percentage of tire failures, ignoring the cost in human lives. Kaizaki said, "we don't think our tires are that bad. Virtually all the tires we're recalling…are good" (Dvorak & Williams, 2000). Ono summarized the outcome of the initial dispute: "We know that we have been slow in responding to public concerns, that we underestimated the intensity of the situation, and that we have been too focused on internal details" ("Firestone Spreads," 2000).

International Implications

During the crisis and in the press, the issue of Japanese-American relations remained minimal. As mentioned above, there were signs the Japanese did not consider issues associated with Firestone to be their problem. However, their solution to the crisis indirectly pointed to cultural conflicts: according to Kaizaki, the fault ultimately lay with Bridgestone for failing to sufficiently "Bridgestone-ise" its American subsidiary ("Bridgestone to Take," 2000). Concretely, this referred to a lack of quality control in the United States, and that advisers would need to be sent to the United States to reform production operations, raising them to Japanese standards (Kranhold & Power, 2000). "President Yoichiro Kaizaki, in a press conference here Monday, admitted that there has been a disparity in quality control standards between the parent and Bridgestone/Firestone, because the US unit has been operated under a policy of localizing itself as an American firm" ("Bridgestone to Unify," 2000).

On the US side, there was no mention of the Japanese connection either. Occasionally, the press mentioned that the US government viewed Firestone's breaking of the 1994–96 strike at the Decatur plant as detrimental, resulting in the personal involvement of the US president at the time, Bill Clinton. This was portrayed as an example of Japanese attitudes toward unions (Merrick, 2000). At times, isolated statements were reported as well, such as one by Gerald Kerr, the marketing director for Firestone during the 1978 recall: "I think the problem is more severe this time. The publicity is worse. The heritage is diffused with Japanese ownership" (Russell & Lin-Fisher, 2000). In general, however, international implications related to the companies' dispute were reflected in their actions rather than public statements.

For these reasons, it would be appropriate to more closely examine ethical implications related to the actions of Ford and Firestone/Bridgestone, more specifically, the ways differences in cultural values, both individually and at the company levels, contributed to this crisis. An examination of this type can serve as a starting point to understand the need for commonly agreed upon principles in international and cross-cultural engineering and business environments:

• If this incident had occurred in the 2010s, how might social media have affected the public's opinion of the companies and the tire recall?
• Do you think companies should be required by law to test products and release test results to the public, regardless of profit motives? Why or why not?

EXERCISE ONE—FORD AND FIRESTONE/BRIDGESTONE (PART ONE)

Before moving on to a discussion of basic ethical principles for global engineering, complete the case study procedure with regard to the case of Ford and Firestone/Bridgestone. Remember, these steps are as follows:

1. Identify ethical issues.
2. Narrow your focus.
3. Determine relevant facts.
4. Make reasonable assumptions.
5. Undertake definitional clarification.
6. Conduct ethical analysis.
7. Review the process.
8. Resolve the issue.
9. Identify practical constraints.
10. Avoid ethical problems.

4.1 THE PRINCIPLES EXPLAINED: ENGINEERING AND JUSTIFICATION[56]

As was discussed in Chapter 1, traditional ethical theories seek to develop one—or a few—principles on which to base all ethical judgments. The approach here differs insofar it begins with the concrete activity of engineering itself. Based

---

56. Materials in this chapter previously appeared in Luegenbiehl (2010a) and (2010b).

on a concept of engineering as a potentially universal discipline, obligations and rights are deduced that would apply to practicing engineers. These principles need to be somewhat flexible in their application, since none can be considered either completely determined or absolute. Rather, these principles should be conceived in and through the process of their application to specific instances—in other words, in conjunction with the case study procedure.

Nevertheless, these principles provide a general framework on which to base ethical judgments in engineering. This approach is similar to the development of codes of ethics for engineers, such as those promulgated by engineering societies throughout the world, as discussed in the previous chapter. However, those codes are delivered as finished products without justification for the individual entries included therein, thus appearing as instruments of external authority, similar in nature to laws. As with laws, these codes are ultimately political instruments based on compromise. Instead, here readers are guided through the process of justifying the adoption of specific principles, thereby having a rational basis for following these principles.

4.2 JUSTIFICATION OF THE PRINCIPLES: ENGINEERING ACTIVITIES

In formulating basic ethical principles for global engineering, an emphasis on any particular, localized perspective should be avoided. Nonetheless, it is necessary to find a foundation on which these principles can be based. This basis should be the nature of engineering itself, an activity-based foundation. In the field of applied ethics, this has become a more common approach. Many texts in medical ethics, for instance, now utilize foundational principles such as *beneficence*, *autonomy*, and *justice* as central to medical practice. These texts tend to take such principles as empirically given, however, as reflecting actual medical practice.[57] Without uniformly standard engineering practices to act as guides, here it is necessary to formulate these principles at a foundational level.

4.3 THE NATURE OF ENGINEERING: VALUE, ARTIFACTS, AND DESIGN

To determine a set of ethical principles for engineers, it is necessary to define "engineering." Historically, a variety of definitions have been proposed. Among these, perhaps the biggest contrast is the emphasis placed on beneficence to humanity—the extent to which engineering should be conceived as either helping humanity or being value neutral. As discussed in Chapter 1, the approach taken here is that engineering should not leave the world less well off than it was before the intervention of engineering activities. This is, thus, a limited value perspective on engineering: although not all engineering needs to have a

---

57. With regard to the field of biomedical ethics, see Beauchamp and Childress (2001).

socially constructive purpose, limits are set on the legitimate activities of engineering. Hence, projects in engineering need not necessarily improve people's lives, although they should leave them no worse off. For the moment, this allows for the avoidance of culturally based conceptions of "benefit" and "harm." That different cultures evaluate different outcomes as more and less beneficial or harmful is a natural consequence of different cultural values, to which we return in Chapter 7.[58]

In defining engineering, another relevant concern is the common public understanding of engineering as being exclusively concerned with making things, the creation of artifacts. This conception is likely linked to engineering's craft tradition. Typically, it is less relevant to the contemporary occupational activities of engineers, which include processes rather than simply products.

Finally, some would argue that engineering is centrally—and perhaps exclusively—concerned with design. While design is certainly a core activity of engineering, it does not fully capture the wide variety of activities in which engineers typically engage. Numerous attempts have been made to segregate engineers into different hierarchies, separating "true" engineers from other types. The question "who are engineers?" is, thus, integrally connected to the nature of engineering, and the answer to this question varies in different countries. Here, for a global context, the definition must thus be somewhat tentative and open to criticism. As was mentioned in Chapter 1, engineering will be defined as *the transformation of the natural world, using scientific principles and mathematics, in order to achieve some desired practical end.*

This is a relatively broad definition, which attempts to capture the widest possible variety of activities in which engineers could potentially be engaged. It is also somewhat value neutral, highlighting the fact that engineering serves practical, desired human ends, while leaving open what these are. This definition also clearly reflects the modern scientific foundation of engineering, rather than the craft tradition, however, which itself involves a value judgment discussed below.[59]

4.4 DERIVING THE PRINCIPLES

With the above understanding of engineering—coupled with the use of reason, case study analysis, and other assumptions discussed in the Chapter 1—one is in a position to examine the principles of ethical behavior that follow. Further examining specific cases within engineering ethics allows for the development of these initial proposals. At this point, using reason, the main concern is determining what makes certain actions appropriate or inappropriate, given the above definition of engineering. The following are elements to keep in mind:

---

58. For an example of different cultural values specific to the work place, see, for example, Shalom (1999).
59. Concerning different definitions of "engineering," see, for example, Dider (2010).

the natural world should be left no worse off as a result of the transformation of engineering, and the costs incurred in this process should not be catastrophic. In addition, when speaking of costs or benefits, the ultimate concern is with the potential impact on human lives, as was stipulated in relation to the previous definition of "ethics." In other words, the appropriate, ethical application of engineers' abilities will be determined in relation to the transformation of the natural world.

The final principles will be somewhat general and open to interpretation. That is, after all, the nature of principles, which, at times, can make them less useful in application than might be desired. This is another reason, however, that readers should practice employing the principles to analyze specific situations, thereby clarifying the nature of the principles for themselves.

4.5   INTRODUCTION TO THE PRINCIPLES: BASED ON PUBLIC SAFETY

The greatest cost an individual can bear is the loss of life or significant injury. Some have argued that life is of infinite value and that, therefore, social benefits cannot be measured against potential losses of life. In practice, however, human life is assigned a value, to different degrees in different cultures. Many insurance companies, for example, determine rates based on, among other factors, one's nationality. However, the importance of the former claim is clear: human life is of very high value. Therefore, actions that risk lives counter potential benefits that result from the introduction of technologies. If engineers are to fulfill their duty not to leave the world worse off, then they should give greatest consideration to the possible endangerment of human life.

This is not to claim that the introduction of technologies potentially harmful to human beings is unjustifiable, since that would imply no degree of risk is justifiable: since nothing is 100% safe,[60] the application of such a principle would eliminate the permissibility of introducing any type of technology. Rather, engineers should give great weight to such risks. This priority of safety is in agreement with statements found in many codes of engineering ethics, for example, that of the National Society for Professional Engineers: "Engineers shall hold paramount the safety, health, and welfare of the public in the performance of their professional duties."[61]

Given that engineers have knowledge and expertise concerning technology unavailable to the general public, one of the responsibilities that follows from their roles as engineers would be the protection of more ignorant individuals from potential dangers. One could reasonably claim that responsibilities follow from knowledge. Taking a commonsense example, if one knows that an

---

60. See Chapter 5 for more on the nature of objective and subjective safety.
61. See the National Society for Professional Engineers' complete code of ethics at its website: http://www.nspe.org/resources/ethics/code-ethics.

individual is about to murder someone, that knowledge generates a duty to warn the person—or take another appropriate action—a duty one would not have if he or she lacked that knowledge. The first basic ethical principle for global engineering can thus be stated as follows:

4.5.1 Public Safety: Engineers Should Endeavor, Based on Their Expertise, to Keep Members of the Public Safe From Serious Negative Consequences Resulting From Their Development and Implementation of Technology

- Which engineering projects benefit our lives but might also endanger public safety? Explain your answer.
- Do you believe human life is of infinite value? Why or why not?

Accepting this principle, a number of others follow. Although the assertion of rights has, perhaps, gone too far in some Western societies, respect for human rights has, at this point, been firmly established on the global level, for example, in the UN convention on human rights. Those asserted most fundamentally as human rights are the protection of life and fulfillment of conditions necessary for the continued existence of life, for instance, food, shelter, education, and just treatment.[62]

For engineers, a duty based on safety—protecting human beings from physical harm—is thus the demand that engineers not undercut conditions necessary for the maintenance of viable human life. Expecting that engineers be responsible for the positive promotion of all human rights, however, would be too much. If that is a positive duty, then it is a duty that belongs to governments and/or the general public. The duty of engineers in relation to rights is, thus, a limited one. Specifically, engineers should not cause the violation of rights through their actions—a duty to respect human rights in carrying out engineering activities and the ability to refuse to participate in engineering activities that threaten such rights. The second basic ethical principle for global engineering then reads as follows:

4.5.2 Human Rights: As a Result of Their Work With Technology, Engineers Should Endeavor to Ensure That Fundamental Human Rights are Not Negatively Impacted

In terms of a global concern, preserving the environment has been a relatively recent phenomenon. Nevertheless, engineers play a major role in sustainable global development, perhaps an insufficiently recognized one.[63] If the environment is not adequately sustained, then human life will clearly be endangered, and a

---

62. The derivation of specific human rights is beyond the scope of this chapter but will be further considered in Chapter 11.
63. Regarding the central role of engineers to sustainable global development, see, for example, Mihelcic, Philips, and Watkins (2006).

similar claim could be made with regard to the destruction of biological diversity. Thus, a third basic ethical principle for global engineering follows from the first principle regarding human life and safety, again, however, as a limited duty:

4.5.3 Environmental Protection: Engineers Should Endeavor to Avoid Damage to the Environment and Living Beings That Would Result in Serious Negative Consequences, Including Long-Term Ones, to Human Life

If engineering activities are carried out in an incompetent fashion, then these activities could have negative consequences that, again, endanger human life. A fourth basic ethical principle for global engineering thus follows directly:

4.5.4 Competent Performance: Engineers Should Endeavor to Engage Only in Engineering Activities They are Competent to Carry Out

- Why do you think engineers might be unaware of the lives that depend on their competent performance?

An important component of the definition of engineering used here is its basis in science and mathematics. In one sense, the employment of science and mathematics is simply a characteristic of competent engineers. In another sense, this employment has wider implications: using other types of principles and decision-making processes would be inappropriate, not in accordance with appropriate engineering procedures. This claim could be considered somewhat controversial, since it implies that engineers are engaged in illegitimate conflicts of interests when they allow nonengineering considerations to influence their judgments. From the perspective of engineering alone, however, nonengineering considerations are irrelevant. A consideration of broader issues will be put aside for the time being, awaiting a later discussion of the wider contexts in which engineering occurs. For the time being, conceiving of engineering in isolation, the fifth basic ethical principle for global engineering is as follows:

4.5.5 Engineering Decisions: Engineers Should Endeavor to Base Their Engineering Decisions on Scientific Principles and Mathematical Analyses, and Seek to Avoid the Influence of Extraneous Factors

- How would this principle come into conflict with the responsibility engineers have to consider the broader implications of their engineering projects? What are examples of extraneous factors that might actually help an engineer perform his or her tasks? Explain your answer.

The fifth principle in this section concerns the direct relationship of engineers to the public. As mentioned before, engineering is a rather esoteric activity, in the sense that much of what engineers do is opaque to the public. However, as justifications for the above principles have shown, engineers cannot take sole responsibility for all the consequences that result from their actions. Although engineers should be worthy of trust, their direct responsibility only relates to engineering aspects of their actions. To begin to fit this responsibility into a larger context, communication with others is necessary. This communication must be of a nature that others can make competent decisions. The sixth basic ethical principle for global engineering is thus as follows:

4.5.6 Truthful Disclosure: Engineers Should Endeavor to Keep the Public Informed of Their Decisions, Which Have the Potential to Seriously Affect the Public, and to be Truthful and Complete in Their Disclosures

- Should engineers act in a completely transparent manner with regard to the public in every situation? Why or why not?

By now, readers should have reflected on the natures of the basic ethical principles for global engineering listed above, and they should be able to apply them to engineering contexts. These principles form a basis—although not an absolute one—for making ethical judgments from the perspective of global engineering. The above principles are based on the nature of engineering itself. At present, they are not based on other considerations that could also be said to legitimately affect the decisions of engineers. This becomes especially apparent in considering the business contexts of global engineering, discussions of which occur in Chapter 6. For this and other reasons, additional principles will be introduced in later chapters, as they become relevant:

- List two more principles you believe would be appropriately included with the principles above. Explain the bases of these principles/their relations to engineering.

EXERCISE TWO—FORD AND FIRESTONE/BRIDGESTONE (PART TWO)

Having been introduced to the first 6 ethical principles for global engineering, readers should return to the work they did on the Ford and Firestone/Bridgestone case above. Redo step 6 of the case study procedure, conducting ethical analysis. Review the first 6 ethical principles for global engineering and decide which of them apply to the most important ethical issue you chose in step 2 of the case study procedure. If conflicts exist between the principles you have identified, then explain the nature of these conflicts, which principle(s) should take precedence, and why you think this is the case. Additionally, if you think other important principles apply to this issue, which have not yet been discussed, then list these principles and provide a brief explanation of why you think they are important to the issue under consideration.

CASE STUDY TWO—DEVELOPMENT AND ITS BROADER CONTEXTS: COAL MINING AND ENERGY, AND THE WEST-EAST PIPELINE IN CHINA

As mentioned before, engineering should not be conceived as value neutral: if engineers did not, on the whole, make the world a better place—or left it worse off—then there would be no need for them. People simply would not want the things for which engineers are responsible, and engineers would not have jobs. Very rarely, however, does the implementation of technologies only have positive consequences. As with most decisions, engineering can have both positive and negative consequences. The task then becomes one of identifying and weighing costs and benefits associated with the introduction of technologies and making decisions based on maximizing anticipated benefits and minimizing potential costs. Although all people face such decisions, in developing countries such decisions can be especially acute.

Governments are faced with providing citizens with resources for living good lives, and technologies play a central role in this process. From food, shelter, energy, and infrastructure to education, jobs, health care, and entertainment, technologies assist in providing resources necessary to flourish. Attendant harms can interfere with the enjoyment of these benefits: adequate food resources might require greater agricultural efficiency, potentially disrupting people's ways of life—where they live and how they work; adequate housing could demand the production and processing of raw materials and the development of previously uninhabited areas that disrupt local environments.

A first step in making responsible decisions with regard to such issues is the recognition that trade-offs exist: at times, many options are less than ideal, and decisions have consequences that result in harms in addition to benefits. A second step involves the exploration of such situations: examining options in terms of the broader social and technological contexts in which they occur helps to better understand these options and, therefore, make better decisions. Toward these ends, the following considers the wide-ranging benefits; harms; and social, political, and economic contexts of development within contemporary China.

Chinese Development: Unparalleled?

In the last 30 years, China has developed more and more quickly than any society in human history, with the largest population, employing technology on a never-before-seen scale. This development is reflected in and fostered by government policies: since China's opening in the late 1970s and early 1980s, political and economic policies have encouraged more autonomous action and personal initiative. Provincial governments have been given more leeway to administer in manners conducive to addressing and benefiting regional differences, and—before having almost completely dismantled the agricultural commune system—farmers were allowed to sell surplus at market price, and many major state-owned enterprises instated executive compensation programs to incentivize profitable performance.[64] The consequences of these policies have been ambiguous.

On the one hand, they have spurred rapid development and raised the standard of living for millions of Chinese. On the other hand, these policies have resulted in an increase in corruption and decrease in public safety.[65] For their revenue streams, provincial governments largely rely on selling land lease rights and taxing local industries. This has resulted in allegedly unfair land grabs, forcing persons from their homes. Coupled with the potential for making tremendous profits in China, conflicts of interests exist between various responsibilities associated with government offices. More recently, these problems have resulted in a greater awareness regarding the nature of and need for individual and collective responsibility.

In these and similar circumstances—given the communal costs associated with individual gains—increased attention has been given to the way the actions of individuals affect the whole. A wide-reaching and ongoing crackdown on corruption has been underway in China since Xi JinPing assumed power, for instance. Central to the media discourse covering such cases is the way the individuals involved have acted irresponsibly in the pursuit of selfish gains at the expense of public wellbeing. These individuals are harshly criticized and made to take responsibility for their actions.[66] In this way then, a greater awareness is arising regarding the nature of and need for responsibility. From the perspective of China's role as a player within sustainable global development, this shift seems promising.

In addition to domestic consequences, the development of China has consequences for the rest of the world, both positive and negative. On the one hand, the development of China has contributed significantly to the global economy, resulting in the formation of new economic and political partnerships. On the other hand, its development severely strains natural resources and contributes significantly to pollution on a global scale.[67]

Defending against charges of this type, many point out that China is still a developing nation, facing challenges different from those of developed countries. Specifically, insofar as industrial modernization requires tremendous resources, a high amount of initial pollution and waste can be expected. These are undoubtedly costs but—the argument runs—costs outweighed by the benefits of development. Further, Western countries have produced large amounts of pollution during periods in their development comparable to China at present. Even today, for example, the United States uses considerably more energy per capita than China. Implied in this defense is the view that development in China mirrors that of the West.

---

64. See both McGregor (2010) and Bell (2015) for more on these policies.
65. For readable accounts of changes in these policies, see Kissinger (2012), Bell (2015), and McGregor (2010). Regarding both their positive and negative effects, see Ma and William (2014).
66. Rather than genuinely rooting out corruption, some have claimed this crackdown aims as stifling political dissent within the Chinese Communist Party.
67. In this regard, see, for example, Ma and William (2014), Muller (2008), Q&A (2013), and China's Pollution (2014).

Insofar as Western nations have already gone through a process of development ultimately benefitting their citizens, taking China to task for undergoing a similar process at present would be hypocritical. Establishing an analogy between the development of China today and the West in the past serves to absolve China of responsibility for the negative consequences of its development. However, development in China seems different from that of the West in several regards.

Rather than the slow, steady implementation of technology—such that this implementation could be stopped or adjusted to account for its consequences—in China's development, the implementation of technology has been much faster, such that little time exists to stop or adjust for its consequences. Assuming engineering is accurately understood as a kind of "social experimentation"—where technology is employed to address concrete problems to make people's lives better—in the case of China's development, this experiment is epic.[68] Further, whereas the consequences of the West's development were relatively local and at the time poorly understood, the consequences of China's development at present are much more global and better understood.

The development of England, Germany, and the United States, for example, took place at a time when the world was less economically and politically connected than today, where the consequences of their development were relatively confined to these countries. Although it would certainly be a mistake to overlook the role played by the ambitions of Western imperialism in shaping development abroad, transportation and communication technologies were such as to put a limit on the consequences of Western development. This is not true today.

At a time when not only China but also India—the two most populous countries in the world—are poised for increased development over the next 50 years, both the positive and negative consequences of the development of any one country are sure to affect those of others.[69] These differences highlight China's importance as a rising world power. Whereas the negative consequences of development in the West were not only unclear but also relatively local, this is not the case with China today.

Although the development of China is different from that of Western countries, it should not be conceived as uniquely Chinese. Insofar as the challenges China faces in its development are similar to those faced by other developing countries—and the world is different enough today from that of Western modernization that such development provides little in the way of models—the development of China can serve as a test case for other developing countries, involving technologies and the contexts for their implementation.[70]

---

68. For more on engineering as social experimentation, see Martin and Schinzinger (2010) and discussions in Chapter 5 on safety.
69. Regarding the relationship between China, India, and Bangladesh in terms of water resources, for example, see Hukil (2013).
70. Concerning the way China's economic development has provided a model for that of other countries, see Kurlantzick (2013), and regarding its political developments, see Bell (2015).

As has been stressed throughout this text, the employment of technology never exists in a vacuum: engineering is itself thoroughly value-laden and exists within social, political, and economic contexts. To a large extent, these contexts determine the natures of benefits and costs associated with technologies. This is especially clear in China with regard to its power industry.

As mentioned above, development requires tremendous energy: power is expended to lay the groundwork for and develop industries and sustain growth. These initial periods of expenditure can and have resulted in environmental destruction and social disruption. To better understand the nature of benefits and costs associated with decisions to employ technology—as well as the social, political, and economic contexts in which these decisions occur—the cases of coal mining and energy and the West-East Pipeline in China are instructive.

Coal Mining and Energy, Public Welfare, Economic Benefits, and Energy Security: Conflicting Interests?

From mine collapses to explosions and sinkholes, coal mining in China is notoriously unsafe.[71] According to the Chinese State Administration of Work Safety, 52,607 people died in mining accidents between 2001 and 2013; 1049 people were killed in 2013 alone (China Mine, 2014). In addition to immediate deaths from mining, various long-term environmental consequences result from the use of coal mining and energy, including air, water, and soil pollution. Despite these negatives effects, China continues to rely on coal to meet its energy needs. China now produces and consumes more coal than any other country in the world, more "than the United States, Europe and Japan combined" (Michieka, Fletcher, & Burnett, 2012). Negatively impacting public welfare not only today but also for generations to come, coal mining and energy in China have tremendous environmental and health costs.[72] Given the nature of trade-offs involving decisions about technologies, one would expect the engineering, social, economic, and political benefits resulting from coal mining and energy to be greater than these costs.

Despite major government investments in clean energy technologies, neither solar panels nor windmills are cost effective enough at present to avoid China's reliance on fossil fuels to meet its current and projected future energy needs.[73] Also, versus fossil fuels such as oil and natural gas, China has an abundance of coal. Given this abundance, coal energy is relatively cheap.[74] This is important since, despite its incredible growth, China is a developing country with a per capita GDP of approximately 6000 USD (Report, 2013). From a consumer

---

71. See China Mine (2014) for more information regarding some of the worst mining disasters in China.
72. See Michieka et al. (2012) for an in-depth account of the various costs of coal energy in China.
73. For an entertaining, very readable account of problems involving clean energy technologies from the perspective of physics, see Muller (2008).
74. See Forsythe (2014) concerning differences in energy policies between China and the United States.

perspective, Chinese people are relatively poor and incapable of paying high-energy costs. From an industrial perspective, this is even more important. China's growth has been primarily export-driven and industrial (Yue, 2008). To fuel this growth, China has relied on coal energy and energy imports such as oil and natural gas, which come primarily from the Middle East, Africa, and Russia. With regard to China's commitment to a uniquely "peaceful rise," securing energy sources from abroad is problematic.

Historically, the economic growth of nations has coincided with military aggression. For these reasons, China's rapid development has alarmed some: assuming the trend holds, one would expect China's economic growth to coincide with increased militarization.[75] The government in Beijing has, thus, sought to assure the international community that its domestic growth poses no threat to either regional or international peace. As a result, China is at times in an awkward position with respect to competing demands for energy imports to fuel its domestic growth and political pressure to act in a manner conducive to international agendas. Its energy policy is, therefore, somewhat constricted. Actual or perceived aggression in securing its energy supply is sure to elicit international responses.[76]

Hence, although China relies on imports to satisfy its energy demands, doing so places it in a somewhat precarious position with respect to international relations. To avoid potential conflicts, as much as possible, China relies on domestic energy sources. Its dependence on coal mining and energy can, thus, be understood in cost-benefit terms—accepting costs in public and environmental welfare because of social, economic, and political benefits.[77] To further alleviate these costs, however, China has invested significantly in the development of other energy resources, especially natural gas.

Gas Energy and the West-East Pipeline, Environmental Welfare, Economic Development, and Urban Bias: Converging Agendas?

Natural gas burns cleaner than coal and yields more energy. For this reason, China has invested considerably in projects to develop natural gas reserves. One such project is the West-East Pipeline (henceforth WEP), transporting natural gas across China, from the Xinjiang autonomous region in the west to the Shanghai metropolitan area in the east. As with coal mining and energy, this project is related to social, economic, and political conditions in China that extend beyond the domain of technical know-how alone. The WEP highlights costs and benefits associated with the employment of technologies and, therefore, the value dimensions of engineering.

---

75. See, for example, the influential account in Huntington (1996).
76. Regarding these issues and their relation to China's peaceful rise, see Yue (2008), Kissinger (2012), and Huntington (1996).
77. For a similar analysis concerning trade-offs in energy security and public safety with regard to the nuclear power industry in Japan, see Luegenbiehl (2009).

Approved by the Chinese State Council in early 2000, the WEP is a core project in China's Western Development Drive and was conceived as serving two main objectives. First, as mentioned above, it would develop natural gas reserves within China, not only increasing energy supply but also decreasing China's dependence on coal energy and natural gas imports to meet domestic energy demands. Second, it would contribute to the development and integration of western, inland regions of China.

Versus eastern coastal cities such as Beijing, Shanghai, and Shenzhen—which are more developed and have higher standards of living—western regions have been less developed and more isolated. As the WEP would run through these regions, the construction and maintenance of the pipeline would contribute to their development through the creation of jobs and by spurring local economies. As with coal mining and energy in China, the results of this project have been mixed.

Upon completion, the WEP is slated to have "an annual delivery capacity of 77 billion cubic meters" of natural gas ("Overview," 2013). With the first phase complete, the pipeline now sends over 2 million cubic meters of natural gas per hour, and the total energy supply would be equivalent to 200 million tons of coal (Liu, 2015). The first phase alone has involved more than 120 cities and 3000 companies (Liu, 2015), and the regions through which the pipeline passes have experienced tremendous economic growth. At the same time, however, the construction and maintenance of the WEP have not benefitted equally all groups involved.

In addition to the Han, China has 55 ethnic minority groups, primarily located in the west and southwest. Although the western regions involved in the construction of the WEP benefitted overall, "nearly all of the equipment and skilled personnel used to develop the Tarim oil basin came from outside Xinjiang" (Herd, 2010). Uyghurs—members of the native Xinjiang minority—make less than their migrant Han counterparts (Wu & Xi, 2013). Some of the most dangerous and menial positions earn the lowest pay, such as line attendants—responsible for maintaining hundreds of kilometers along the pipeline in potentially dangerous conditions (Yu, 2009, pp. 54–55). Working in high-risk environments, members of the emergency crew, for instance, were paid 60 RMB per day, less than 10 USD (Yu, 2009, pp. 151–152).

Some have claimed "the massive investment" associated with the WEP "has mainly benefited state-owned companies" in China (Richburg, 2010). In 2014, for example, 80 billion stocks in the WEP were sold, primarily benefitting the China National Petroleum Corporation—a state-owned enterprise responsible for the WEP—and private capital investors. Although investment from the WEP has helped the regions through which it passes, there are reasons to doubt the long-term sustainability of these benefits.

Drawing an analogy, the Yiqikelike oil field—also in Xinjiang—operated for 30 years. With the oil reserves depleted, however, today, the town stands in ruin: "collapsed auditoriums, schools, and refineries; rusty oil wells; murals and slogans of the past age" (Yu, 2009, pp. 36–37). Some worry the same eventual fate awaits those regions benefitting from the WEP today:

- Reflect on and explain two situations that result from conflicting values with which you are familiar, from either your own life or news media. List the relevant values and decide which ones are given priority. How do/would you decide which values should be given priority?
- Find out the largest source of energy in your country—for example, coal, oil, natural gas, nuclear, or solar—listing and explaining benefits and costs associated with this source.

EXERCISE THREE—DEVELOPMENT AND ITS BROADER CONTEXTS

Complete the case study procedure on Development and its Broader Contexts using the first 6 ethical principles for global engineering. If you think other important principles apply to this issue, which have not yet been discussed, then list these principles and provide a brief explanation of why you think they are important to the issue under consideration.

4.6 SUMMARY

As the case of Ford and Firestone/Bridgestone illustrates, given the increasingly international and cross-cultural nature of engineering and business environments, the need exists for broad but commonly agreed upon principles of global engineering ethics. However, as discussed in Chapter 1, different persons and peoples subscribe to and are influenced by different cultural and social values, which present difficulties in formulating commonly agreed upon ethical principles for engineering in global contexts. In addressing these difficulties, a number of assumptions were made in Chapter 1, one of which concerns the need to begin with the universal nature of engineering to arrive at principles of engineering ethics and the priority of public safety. As opposed to pregiven engineering ethical codes—which can appear simply as an imposition by external authority—the approach taken here consists in deriving principles for engineering ethics in global contexts through the use of reason, where the reader can follow along, better understanding, justifying, and ultimately employing these principles. Based on and derived from the nature of engineering itself and the primacy of public safety, these first 6 principles govern engineers' relationships with members of the public and their safety, fundamental human rights, the environment and biological diversity, engineers' competences in engineering activities, the use of scientific and mathematical principles in engineering activities, and communication with the public. This is by no means a comprehensive list of principles governing the ethical behaviors of engineers, and further principles are derived, formulated, and justified in the context of discussions of relevant contents in later chapters. The case Development and its Broader Contexts begins to show that the activities of engineers always occur in broader contexts, where social, political, and economic factors are also relevant.

REVIEW QUESTIONS

1. In the case of Ford and Firestone/Bridgestone, explain how tire pressure became a major cause of conflict between the two companies.
2. Where did Ford initiate the tire replacement program, and in what manner did they publicize it?
3. In what ways did leadership within Firestone/Bridgestone cause confusion when carrying out the recall?
4. List and describe two differences in the values of Ford and Firestone/Bridgestone that contributed to the tire crisis. How do these differences motivate the need for universal principles of global engineering ethics?
5. Describe four issues that could have been avoided during the Ford and Firestone/Bridgestone case, if a less localized ethical and value system was used by both companies.
6. Describe three central characteristics of the definition of "engineering" outlined above.
7. Explain three reasons that engineering ethical principles should be based on public safety.
8. Why should engineers give significant weight to risks associated with new technologies? Explain how risk is reflected in each of the six global principles described above.
9. List and describe particular challenges facing developing countries and regions.
10. How is development within China similar to that of Western countries? How is it different?
11. List and explain both benefits and costs associated with coal mining and energy in China.
12. How has the WEP both benefitted and hurt Chinese people?
13. List concerns and/or principles relevant to the decision by the Chinese government to pursue the energy policies described in Development and its Broader Contexts.

REFERENCES

A Message From John Lampe. (2000). *Bridgestone/Firestone Corporate News*. October 10. http://mirror.bridgestone-firestone.com/recall/indexrcnav.html.

Aeppel, T. (2000a). Bridgestone consolidates 16 tire units into four to provide better oversight. *The Wall Street Journal*, A6. October 19.

Aeppel, T. (2000b). Bridgestone unit's CEO says firm hasn't found defect. *The Wall Street Journal*, A4. October 27.

Aeppel, T. (2001). Firestone has been here before. *The Wall Street Journal*, A16. September 6.

Aeppel, T., Power, S., & Geyelin, M. (2000a). Firestone breaks with ford over tire pressure. *The Wall Street Journal*, A3 & A10. September 22.

Aeppel, T., Power, S., & White, J. (2000b). Ford, firestone decide to work together to find explanation for tire problems. *The Wall Street Journal*, A3. December 1.

Analysis: Bridgestone Caught Asleep at Wheel. (2000b). *Nikkei Net Interactive*. September 8. http://www.nni.nikkei...TNKS/Search/Nni20000908DNBN908M.htm.

Beauchamp, T., & Childress, J. 2001. *Principles of biomedical ethics*. Oxford: Oxford University Press.

Bell, D. (2015). *The China model: Political meritocracy and the limits of democracy*. Princeton, NJ: Princeton University Press.

Bridgestone Fiddles as Firestone Burns. (2000). *Nikkei Net Interactive*. September 8. http://www.nni.nikkei...tnks/Search/Nni20000908DNBN908C.htm.

Bridgestone Forecasts Loss After U.S. Unit Recalls Tires. (2000). *The Japan Times Online*. August 11. http://www.japantimes.co.jp/cgi-bin/getarticle.pl5?nb20000811a3.htm.

Bridgestone Names New President. (2001). *Bridgestone/Firestone Corporate News*. January 11. http://www.bridgestone-firestone.com/news/corporate/news/010111a.html.

Bridgestone President Denies Cover-Up of Faulty Tires. (2000). *Nikkei Net Interactive*. September 19. http://www.nni.nikkei...tnks/Search/Nni20000918D18JF137.htm.

Bridgestone to Take the Reins at US Unit. (2000). *Financial Times*. September 12. 36.

Bridgestone to Unify Quality Control Standard in Japan, U.S. (2000). *Jiji Press English News Service*. September 11. http://proquest.umi.com/pqdweb?TS=975509...=1&Dtp=1&Did=000000059840720&Mtd=1&Fm.

Bridgestone Will Back Firestone, Kaizaki Says. (2000). *The Japan Times Online*. September 12. http://www.japantimes.co.jp/news/2000/09/12/business/bridgestone-will-back-firestone-kaizaki-says/#.WAHLbjJh2CQ.

Bridgestone/Firestone CEO Provides Update on Investigation. (2000). *Bridgestone/Firestone Corporate News*. August 23. http://mirror.bridgestone-firestone.com/news/corporate/news/00823a.htm.

China Mine Disaster Watch. (2014). June 1 Retrieved from http://www.usmra.com/chinatable.htm.

China's Pollution Altering US Weather, Claim Scientists (2014). *South China Morning Post*. April 17. http://www.scmp.com/news/china/article/1485843/chinas-pollution-altering-us-weather-claim-scientists.

Dider, C. (2010). Professional ethics without a profession: A french view of engineering ethics. In I. Poel & D. Goldberg (Eds.), *Philosophy and engineering. An emerging agenda*. Dordrecht: Springer.

Dobbyn, T. (2000). Ford says it properly tested firestone tires. *The Safety Forum News*. September 20. http://www.safetyforum.com/cgi-bin/sn_search.cgi?ID=000582.

Dreazen, Y. (2000). Deaths continue during firestone recall. *The Wall Street Journal*, A3 & A7. September 14.

Dvorak, P., & Williams, M. (2000). Bridgestone may restaff its U.S. unit. *The Wall Street Journal*, A21 & A23. September 19.

Dvorak, P., & Williams, M. (2001). Embattled CEO of Japan's Bridgestone resigns. *The Wall Street Journal*, A6. January 12.

Dvorak, P., & Zau, T. (2000). To grasp the tire case, pay a visit to Mr. Seki, a very mild regulator. *The Wall Street Journal*, A1 & A6. September 8.

Eisenberg, D. (2000). Anatomy of a recall. *Time*, 29–32. September 11.

ElBoghdady, D. (2000). Tire probe intensifies. *The Detroit News*, August 7. http://www.detroitnews.com/2000/autos/0008/07/a01-101797.htm.

Firestone Letter to Belo & KHOU Executives. (2000). February 10. http://www.khou.com/news/stories/1290.html.

Firestone Spreads Blame (2000). *CNN.fn.* September 11. http://cnnfn.cnn.com/2000/09/11/companies/bridgestone/html.

Fogarty, T., & Eldridge, E. (2000). Lawmakers fault tire tests, pressure. *USA Today*, B3. September 22.

Ford Authorizes Dealers to Use Non-Firestone Replacements. (2000). *The Safety Forum News.* August 11. http://www.safetyforum.com/cgi-bin/sn_search.cgi?ID=000540.

Ford Motor Company. 2000. *Company news room.* September 6. http://www.ford.com/default.asp?pageid=106&storyid=946.

Forsythe, M. (2014). As U.S. aims for energy independence, China heads the opposite way. *New York Times*, February 13. http://sinosphere.blogs.nytimes.com/2014/02/13/as-u-s-aims-for-energy-independence-china-heads-the-opposite-way/.

Fourth Bridgestone Plant Ups Output. (2000). *The Japan Times Online.* August 30. http://www.japantimes.co.jp/cgi-bin/getarticle.pl5?nb20000830a5.htm.

French, H. (2000). Japan wonders what became of quality control. *The New York Times*, A3 September 9.

Geyelin, M. (2000). Squabbles between Ford and Firestone may hurt their legal defense. *The Wall Street Journal*, B1 & B6 October 9.

Greenwald, J. (2000). Firestone's tire crisis. *Time*, 645 August 21.

Grimaldi, J. (2001). Second Firestone plant faulted. *The Washington Post Online*, August 18. http://www.washington...-dyn/articles/A44774-2000Aug17.html.

Healey, J., & Nathan, S. 2000a. Ford might drop Firestone tires. *USA Today*, August 3. http://www.usatoday.com/money/consumer/autos/mauto719.htm.

Healey, J., & Nathan, S. (2000b). Further scrutiny puts Ford in the hot seat. *USA Today*, B1 September 21.

Herd, R. (2010). A pause in the growth of inequality in China? OECD Economics Department Working Papers, No. 748, OECD Publishing, Paris. http://dx.doi.org/10.1787/5kmlh52r90zs-en.

Hukil, R. (2013). India, China, and Bangladesh: The contentious politics of the Brahmaputra river. *Institute of Peace and Conflict Studies*, March 9 http://www.ipcs.org/article/india/india-china-and-bangladesh-the-contentious-politics-of-the-brahmaputra-3840.html.

Huntington, S. (1996). *The clash of civilizations and the remaking of world order.* New York, NY: Simon and Schuster.

Hyde, J. (2000). Ford Motor Co. raises pressure recommendation for explorer tires. *Terre Haute Tribune-Star*, A3. September 23.

Kiley, D., & Healey, J. (2000). Ford CEO handles tire recall. *USA Today*, August 17. http://www.usatoday.com/money/consumer/autos/mauto765.htm.

Kissinger, H. 2012. *On China.* New York: Penguin Books.

Kranhold, K., & Power, S. (2000). Bridgestone turns to Ketchum to Redo image after tire recall. *The Wall Street Journal*, A4 September 12.

Kunii, I., & Foust, D. (2000). They just don't have a clue how to handle this. *Business Week*, 43. September 18.

Kurlantzick, J. (2013). Why the 'China model' isn't going away: From Bangkok to caracas, Beijing's style of authoritarian capitalism is gaining influence. *The Atlantic*, March 21. http://www.theatlantic.com/china/archive/2013/03/why-the-china-model-isnt-going-away/274237/.

Liu, X. (2015). West-east gas pipeline project first phrase transported 150 billion cubic meters natural gas and benefited 400 million residents. (in Chinese) Xinhuanet. June 21.

Luegenbiehl, H. (2009). Societal values and nuclear power: A case of conflicting priorities. *Humanities and Technology Review*, 28, Fall.

Luegenbiehl, H. (2010a). Principles for cross-cultural engineering ethics. *2010 American Society for Engineering Education Global Colloquium*, Singpore, Oct.

Luegenbiehl, H. (2010b). Ethical principles for engineers. In I. Van de Poel & D. Goldberg (Eds.), *Philosophy and engineering. An emerging agenda* (pp. 147–159). Berlin: Springer.

Ma, D., & William, A. 2014. *In line behind a billion people: How scarcity will define China's ascent in the next decade*. Upper Saddle River, NJ: FT Press.

Martin, M., & Schinzinger, R. 2010. *Introduction to engineering ethics* (2nd ed.). New York, NY: McGraw Hill.

Mayer, C., & Swoboda, F. (2000). I come…to apologize. *The Washington Post Online*, September 7. http://www.washington.p-dyn/articles/A23761-2000Sep6.html.

McGregor, R. 2010. *The party: The secret world of China's communist rulers*. New York, NY: Harper Perennial.

Merrick, A. (2000). Bridgestone tire issue clouds labor negotiations. *The Wall Street Journal*, A4. September 1.

Michieka, N., Fletcher, J., & Burnett, W. (2012). *The cost of energy: The environmental effects of coal production in China*. In: 31st USAEE/IAEE conference, November 4–7. http://www.usaee.org/usaee2012/submissions/OnlineProceedings/The%20cost%20of%20energy_env_effects%20of%20coal_Burnett.pdf..

Mihelcic, J., Philips, L., Watkins, D., Jr. (2006). Integrating a global perspective into education and research: Engineering international sustainable development. *Environmental Engineering Science, 23*(3), 426–438.

Miller, K. (2000). Attorneys question Bridgestone/Firestone chief. *Terre Haute Tribune-Star*. October 10. A3.

Miller, K. (2000). Firestone's top executive steps down. *Terre Haute Tribune-Star*. October 11. A4.

Missteps Cost Bridgestone Dearly. (2000). *Nikkei Net Interactive*. October 30. http://www.nni.nikkei…tnw/Search/Nni20001030IN108000.htm.

Muller, R. 2008. *Physics for future presidents: The science behind the headlines*. New York, NY: W.W. Norton and Company.

Jac Nasser, President and CEO of Ford Motor Company, Statement in Response to NHTSA Investigation. (2000). Media.ford.com. August 4. http://media.ford.com/article_display.cfm?article_id=5573.

Opinion: Bridgestone Needs New PR Approach for Tire Fiasco. (2000). *Nikkei Net Interactive*. September 18. http://www.nni.nikkei…tnks/Search/Nni20000918DNBN917C.htm.

Overview of the West-East Gas Pipeline Project (2002–2013). (2013). West-east gas pipeline project (2002-2013) Special Report on Social Responsibility. CNPC. http://www.cnpc.com.cn/en/cs2012en/201407/3d2ccb479ad94ef4a6c54ce4d78685fa/files/8440f95e4b454eb082d557b5261d667c.pdf.

Pickler, N. (2000a). Firestone to recall tires. *Associated Press*. August 8. http://www.safetyforum.com/cgi-bin/sn_search.cgi?ID=000503.

Pickler, N. (2000b). CEO apologizes for fatal accidents possibly linked to tires. *Terre Haute Tribune-Star*. September 7. A4.

Pickler, N. (2001). Government hears of 26 more tire deaths. *Terre Haute Tribune-Star*. February 7. A8.

Power, S., & Simison, B. (2000). Bridgestone and Ford give their defense in hearings on handling of tire recall. *The Wall Street Journal*, A3 & A10. September 7.

Q&A: China-Japan Island Row. (2013). *BBC News: Asia*. November 27. http://www.bbc.co.uk/news/world-asia-pacific-11341139.

Report for selected countries and subjects. (2013). International Monetary Fund. http://www.imf.org/external/pubs/ft/weo/2013/02/weodata/.

Richburg, K. (2010). China's push to develop its west hasn't closed income gap with east, critics say. *The Washington Post*. June 29. http://www.washingtonpost.com/wp-dyn/content/article/2010/06/28/AR2010062804979.html.

Russell, J. & Lin-Fisher, B. (2000). Firestone: Can it survive the firestorm of recall. *Terre Haute Tribune-Star*. September 10. A7.

Schneider, G. (2000). Customers find dealers unready for tire recall. *The Washington Post Online*, August 11. http://www.washington...p-dyn/articles/A6534-2000Aug10.html.

Sears Pulling Firestone Tires from Stores. (2000). CNN.com. August 4. http://www.cnn.co...8/04/tiredeaths.sears.ap/index.html.

Shalom, S. (1999). A theory of cultural values and some implications for work. *Applied Psychology*, 48(1), 23–47.

Shirouzu, N., & Aeppel, T. (2000). Firestone says it acted to improve problem tires. *The Wall Street Journal*, A4 &A6. August 18.

Shirouzu, N., & Power, S. (2000). Nasser terms recall a disappointment in long relationship. *The Wall Street Journal*, A3 & A4. September 1.

Simison, R. (2000). For Ford CEO Nasser, damage control is the new 'Job One'. *The Wall Street Journal*, A1 & A8. September 11.

Simison, R., Shirouzu, N., & Aeppel, T. (2000). Ford says it knew of Venezuelan tire failures in 1998. *The Wall Street Journal*, A3 & A8. August 20.

Simison, R., Shirouzu, N., Aeppel, T., & Zaun, T. (2000). Tension between Ford and Firestone mounts amid recall efforts. *The Wall Street Journal*, A1 & A8. August 28.

Skertic, M. (2000). Failing tires carry fatal consequences. *Chicago Sun-Times*, April 30 http://www.suntimes.com/tires/tires1b.html.

Spindle, B. (2000). Japanese traditions protect Bridgestone from threat of raiders. *The Wall Street Journal*, A10. September 7.

Spurgeon, D. (2000). Another Firestone tire is under scrutiny. *The Wall Street Journal*, A3 & A6. August 31.

Spurgeon, D. (2001). State farm researcher's sleuthing helped prompt Firestone recall. *The Wall Street Journal*, B1 & B6. September 1.

Statement by Christine Karbowiak, Vice President, Public Affairs, Bridgestone/Firestone, Inc. (2000). *Bridgestone/Firestone Corporate News*. September 5. http://mirror.bridgestone-firestone.com/news/corporate/news/000905b.htm.

Statement by Gary Crigger. (2000). *Bridgestone/Firestone Corporate News*. August 9. http://mirror.bridgestone-firestone.com/news/corporate/news/00809c.htm.

Tire Failures on Ford SUV's Producing Alarming Number of Crashes, Deaths. (n.d.). *The Safety Forum News*. http://www.safetyforum.com/cgi-bin/sn_search.cgi?ID=000473.

Truby, M. & ElBoghdady, D. (2000). More deaths linked to tires. Detnews.com. August 16. http://detroitnews.com/2000/autos/0008/16/b01-106172.htm.

Wu, X. & Xi, S. (2013). Ethnicity, Migration, and Social stratification in China: Evidence from Xinjiang Uyghur autonomous region. PSC Research Report No. 13-810. November.

Yu, M. (2009). *The cloud and months: Walking on the ground of west-east pipeline (in Chinese)*. Chongqing: Chongqing Publishing Press.

Yue, J. (2008). Peaceful rise of China: Myth or reality? *International Politics*, 45, 439–456.

Zaun, T. (2000). Firestone Finds No Tire Flaw. *Asia Wall Street Journal*, October 30–November 5. 6.

Zaun, T., & Dvorak, P. (2000). Firestone's Japan parent appears anxiety-free despite U.S. recall. *The Wall Street Journal*, A16. September 5.

Zaun, T., Dvorak, P., Shirouzu, S., & Landers, P. (2000). Bridgestone boss has toughness, but is that what crisis demands? *The Wall Street Journal*, A1 & A18. September 12.

Zaun, T., & Shirouzu, N. (2000). Bridgestone may ask U.S. chief to step down. *The Wall Street Journal*, A3. October 3.

Zaun, T., White, J., & Aeppel, T. (2000). Firestone parent will set aside cash for claims. *The Wall Street Journal*, A6. December 6.

Zaun, T., & Woodruff, D. (2000). Bridgestone boost tire production in Japan to support recall. *The Wall Street Journal*, A8. August 30.

Chapter 5

The Prime Responsibility
of Safety

Chapter Objectives

Having read this chapter, completed the included exercises, and answered the
associated questions, readers should be able to

- with reference to the case of Hurricane Katrina, explain the value of and
  difficulty in studying disasters, identify and apply the basic ethical principles
  for global engineering, and identify competing claims made on engineering
  and engineers from the perspective of safety;
- describe the ways engineering can be understood as a kind of "social
  experimentation" and the responsibilities of engineers that follow from this
  analogy;
- explain the natures of and criteria for assessments of objective and subjective
  safety and how these present challenges to engineers, especially in cross-
  cultural contexts;
- with reference to the case of the Uber Rape Scandal, explain some
  responsibilities that engineering and technology firms could be claimed to
  have to the safety of their users.

CASE STUDY ONE—HOW SAFE IS SAFE?: THE CASE OF HURRICANE KATRINA[78]

When large-scale disasters occur, a common response consists in looking for
the causes of such catastrophes—whether human or natural—to assign blame.
As in the case of the Überlingen midair collision, more often than not, no single cause is solely responsible. This makes the assessments of such disasters
fraught with uncertainty and, at times, can result in scapegoating—finding one
party on which to unfairly place all the blame. In retrospect, technologies, decisions, organizations, etc., will be described as "unsafe." A more positive response would be to acknowledge there is always "enough blame to go around,"
focusing instead on assessments of safety and the future, so that similar events
can be prevented.

---

78. Materials in this case study previously appeared in Luegenbiehl (2007).

Global Engineering Ethics. http://dx.doi.org/10.1016/B978-0-12-811218-2.00005-9

Both responses were parts of the public discourse surrounding Hurricane Katrina in 2005, one often to the exclusion of the other. Obviously, however, neither is incompatible with the other and, in fact, both should play a role in forming a total assessment of the situation: without understanding why events occur, it is difficult to learn from them. Generally, if suspicions exist with regard to assigning responsibilities, then they concern the motives of those seeking to blame—diverting attention away from themselves, wanting to beat their opponents, or exhibiting moral superiority against others.

Complexity is a second issue. Given the confluence of various causal conditions, it is difficult and, perhaps, unfair to place all the responsibility on individual actors or institutions. The Independent Levee Investigation Team report on Hurricane Katrina, for example, states the following: "In the end, it is concluded that many things went wrong with the New Orleans flood protection system during Hurricane Katrina, and that the resulting catastrophe had it[s] roots in three main causes: (1) a major natural disaster (the hurricane itself); (2) the poor performance of the flood protection system due to localized engineering failures, questionable judgments, errors, etc., involved in the detailed design, construction, operation and maintenance of the system; and (3) more global "organizational" and institutional problems associated with the governmental and local organizations responsible for the design, construction, operation, maintenance, and funding of the overall flood protection system" (Seed et al., 2005, xviii). While this list explicitly identifies three major causes for the disaster, it implicitly invokes a myriad of individual sources, many of which can only be properly assessed in conjunction with other underlying factors. The task of assigning responsibility would, therefore, be incomplete and somewhat subjective.

In cases of large-scale disasters—when suspicions of motives are combined with the inherent difficulty of assigning responsibility to individuals or organizations—it might seem as though reasons exist for avoiding the ethical assessment of such cases. In learning about ethics, some might argue, it is better to examine clear-cut situations in which ethical questions arise, where it is easier to assign responsibility and arrive at conclusions regarding ethical issues. If the goal of ethics education is to learn lessons from the past for the future, then perhaps simpler is better.

In contradistinction to this line of thought, the position taken here is that—given the increasingly complex nature of engineering environments and diverse actors involved—examples like Katrina are robust and appropriate sources for learning about ethics, revealing important lessons not generally learned from other types of cases. Cases about disasters show the dramatic, at times catastrophic nature of consequences that can result from the confluence of relatively minor, seemingly insignificant decisions. Such circumstances highlight the paramount importance of public safety to engineering ethics and standards to address the nature of safety.

Katrina as a Case for Engineering Ethics

Describing the consequences of Hurricane Katrina, the report of the Independent Levee Investigation Team states the following: "This event resulted in the single most costly catastrophic failure of an engineered system in history. Current damage estimates at the time of this writing (May 22, 2006) are on the order of $100–150 billion in the greater New Orleans area, and the official death count in New Orleans and southern Louisiana at the time of this writing stands at 1293, with an additional 306 deaths in nearby southern Mississippi" (Seed et al., 2005, xviii).

Before assessing this case as one appropriate for engineering ethics, however, the possibility that it would fall outside the scope of engineering ethics should be considered. One might argue, for example, that, as a disaster, Katrina was an "act of God," due primarily to forces of nature rather than human actions, and, thus, that it was ultimately unpreventable. Or one might claim that the origins of the situation lie so far in the past that assignment of responsibility is a fruitless enterprise. Further, one might point to the limited power and authority engineers exercised in the creation and maintenance of the flood prevention system, or that those in New Orleans were responsible for their own fates, having made the decision to reside there. All of these points have been debated, and considering them all in detail would be beyond the present scope. For the purposes here, establishing an engineering ethics dimension with regard to this disaster is enough.

Speaking to this point, Raymond Seed, the head of the Independent Levee Investigation Team and professor of civil engineering, asserted the following on the PBS NewsHour: "The levee system failed in large part because of embedded deficiencies and because safety and reliability were put at risk; they were traded for economic efficiencies" (PBS, 2006). The report on Katrina mentioned the following specifically engineering-related failures:

Inadequate margins of safety. The factor of safety used for the design of the levee system was inappropriately low to protect an urban population. It would have been more suitable for agricultural farmland (Seed et al., 2005, xxiv, pp. 15–10).

An incomplete system. At the time of Hurricane Katrina, sections of the New Orleans levee system were incomplete, leaving gaps in the system (Seed et al., 2005, pp. 8–16). Also, sections were below their design height specifications (US Army, 2006, I.1–2).

A fragmented system. Various local organizations controlled different parts of the levee system. This resulted in incompatible structures with inherent weaknesses at points of transition (Select Bipartisan Committee, 2006, pp. 91–92).

Inadequate design. Failures of some levees could have been prevented with relatively simple measures, such as installing concrete splash pads at the base of the levees to prevent erosion or using T- rather than I-shaped walls (Seed et al., 2005, pp. 8–7). Floodgates could have been installed at the entrance of the New Orleans drainage canals, which did not occur because of the infighting of local agencies (Seed et al., 2005, pp. 8–15).

Failure to learn from others. The Corps of Engineers commissioned studies for the improvement of the levee system, but the results of these studies were ignored in subsequent design decisions (Seed et al., 2005, pp. 8-13). In the Netherlands, a flood control system is used with the ability to protect against a Category 5 hurricane. At landfall, Katrina was a Category 3, and by the time it hit New Orleans, it was even weaker (Seed et al., 2005, pp. 8–13; Graumann et al., 2005, pp. 1–2).

Lack of central control. The fact that no one had overall oversight of the levee system was part of the reason for the failure of the system. Different local agencies were responsible for various sectors, without adequate coordination among them (Seed et al., 2005, pp. 8–5).

The above is by no means a complete list of possible engineering issues related to Hurricane Katrina, but it is sufficient to show that this case is suitable for analysis in terms of engineering ethics. The engineering failures discussed above reflect a general consensus among different reports on Katrina, although differences exist among these regarding assessments of specific responsibilities:

• In which ways is a leadership model based on central control more effective than one in which numerous agencies work on a project? What are some potential downfalls of a central control model?

Studying Katrina

Making the claim that Katrina is a case appropriate for study within engineering ethics is, of course, not the same as making judgments that engineers are—in part or whole—to blame for the consequences of Katrina. However, that is a conclusion at which some might arrive, based on statements such as those made by Carl Strock, lieutenant general and member of the US Army Corps of Engineers, on the PBS NewsHour: "At the end of the day, we have accumulated a level of risk and I don't think we truly understood that. So I think what you'll see is that, before we defer to others on elements that involve engineering decision-making and our ethical responsibilities to ensure what we build is going to serve its purpose, I think you'll see a greater propensity on the part of the Corps of Engineers to stand up and say 'No'" (PBS, 2006).

Instead of simply blaming engineers, using a disaster like Hurricane Katrina as a case from which to learn points to the complexity characteristic of sophisticated ethical analysis. Simply because engineers "put at risk" safety or failed to exercise professional autonomy does not automatically mean they should be blamed for what happened. Precisely because they are not susceptible to simplistic conclusions, large-scale disasters provide perfect cases from which to learn more about engineering ethics. Incidences like Hurricane Katrina should be used to consider a variety of possible interpretations and conclusions, with the understanding that no one conclusion is necessarily the only right one or even the best. Hence, the point of examining disasters in engineering ethics is not to assign blame but to discuss issues. Both the natures of the disaster and sources about Katrina make it a case that provides rich opportunities for discussions.

When Henry Petroski entitled his popular *To Engineer Is Human*, he surely had in mind the saying "to err is human, to forgive divine." Like all other human beings, engineers are far from perfect. However, most strive to do their best. To do so, they must learn from the past. In itself, this is an important lesson to be learned from ethical analysis. As Petroski rightly says, "no disaster need be repeated, for by talking and writing about the mistakes that escape us we learn from them, and by learning from them we can obviate their recurrence" (Petroski, 1985, p. 227).

EXERCISE ONE—HOW SAFE IS SAFE? (PART ONE)

Listed below are sets of ethical issues regarding/questions about the case of Hurricane Katrina. For each set of issues, complete the following steps of the case-study procedure:

1. Three—listing facts from the case relevant to resolving the issues under consideration and any important missing facts.
2. Four—writing out reasonable assumptions one can make regarding these missing facts.
3. Five—clarifying the use of terminology, in the ethical issues, relevant facts, missing facts, or reasonable assumptions.
4. Six—referring back to the basic ethical principles for global engineering described in Chapter 4, list the ones that apply to the issues under consideration here. If apparent conflicts exist between the principles you have listed, then rank them, indicating which principles you think should take precedence and providing short justifications for the hierarchy. Additionally, if you think there are other important principles that should apply—from your personal life or in general—then list these as well, again, providing brief justifications for the relevance of these principles.
5. Seven—going back through the previous steps, see if there are other issues or facts that have been overlooked, concepts that can be clarified, or principles applied.
6. Eight—resolving the issue under consideration/answering the question, provide a brief justification.
7. Nine—looking over the case again, identify and list any constraints that could reasonably excuse either individuals or organizations from the resolution/answer given.

8. Ten—outlining how ethical problems might have been avoided in the first place.

- Given General Strock's description above, was the Corps of Engineers sufficiently autonomous in its decision-making processes? To what extent should engineers be able to exercise their professional autonomy—the right to make decisions based on their specialized, professional knowledge?
- Was preparing for a Category 3 hurricane adequate for a densely populated area prone to hurricanes? What level of risk is acceptable in different kinds of situations?
- Should the issue of "whistle blowing"—bypassing a hierarchical chain of command or exposing information to outside bodies—have arisen for engineers involved in the building of the levee system? Do engineers have specific professional and ethical responsibilities that go beyond ordinary ethical duties?
- Should those warned about the state of the levee system prior to the hurricane have taken greater actions? What is the role of professional organizations and outside parties?
- Did the levee designers have a responsibility to consider the ways different social strata in New Orleans would be affected by major flooding? Should engineers consider the differential consequences of their design decisions on different groups?
- At what point do the risks involved in levee construction outweigh the potential benefits to the local population? Do engineers have a positive duty to do good—beyond the simple duty of avoiding harm?
- Is a military organizational structure suitable for constructing urban civil works that involve the safety of large numbers of people? Can engineers be expected to answer to several different "masters"?

5.1   SAFETY: A SPECIAL CONCERN FOR ENGINEERS— ENGINEERING AS "SOCIAL EXPERIMENTATION"

As the case of Hurricane Katrina shows, modern technology has the incredible power to affect human lives. What might, at times, seem like minor design decisions can bring harms or benefits to millions or even billions of people. Even the powers of contemporary dictators are small compared with the powers of engineers. These powers carry with them a great deal of responsibility. However, this often goes unrecognized, even by the holders of the powers themselves. While the powers of engineers can bring great benefits to the public, they can also bring great harms.

Neglecting appropriate safety considerations in the design of a nuclear reactor, for example, could destroy lives and make areas of land uninhabitable for decades. An emphasis on safety highlights the centrality of ethics in engineering. It means engineers are responsible for not only the technical adequacy of their activities but also the consequences that result from the intended and un-intended—but foreseeable—effects of these activities, insofar as they have the potential to harm the public.

To discuss duties of safety engineers have to the public, Mike Martin and Roland Schinzinger introduce and develop—what has come to be—a well-known analogy between engineers and social experimenters. The authors frame engineering as a kind of social experimentation, showing that specific responsibilities follow for engineers, as they do for scientists performing experiments on human subjects. As with scientific experiments on human subjects, the activities of engineers are carried out—at least in part—in ignorance, where the outcomes of experiments are uncertain: the introduction of new technologies into society can have unknown consequences. Unlike with scientific experiments, however, as experiments, engineering activities lack controls that would act as protections—that is, no control group is established or alternative reality developed—and are carried out on much larger scales. The responsibilities of engineers would, therefore, be even greater than those of scientific experimenters.

In *Introduction to Engineering Ethics*, Martin and Schinzinger define these responsibilities as follows: "(1) A primary obligation to protect the safety of human subjects and respect their right of consent. (2) A constant awareness of the experimental nature of any project, imaginative forecasting of its possible side effects, and a reasonable effort to monitor them. (3) Autonomous, personal involvement in all steps of a project. (4) Accepting accountability for the results of a project" (Martin & Schinzinger, 2010, p. 86). Thus, according to Martin and Schinzinger, the responsibilities of engineers for safety cover the full range of engineering activities. Additionally, they believe these are responsibilities applicable to each individual engineer involved in processes of engineering activities:

● How can the Hurricane Katrina disaster be conceived as a social experiment in relation to engineering, and why is this example particularly important in considering future safety measures?

5.2 THE NATURE OF SAFETY: OBJECTIVE AND SUBJECTIVE

All human beings are concerned with safety. This concern probably stems from universally evolved biological instincts that protect against death and serious injury. One of the easiest ways to keep members of the public from engaging in activities is to convince them that these activities are unsafe. However, it is unclear precisely what "being safe" means.

Safety is strongly associated with freedom from harm, although harm can take many forms. Harm can be physical, psychological, emotional, financial, and so on in nature. Holding engineers responsible for all the potential harms that can result from introducing technologies into society would, clearly, be unjustifiable, although engineers might have contributory responsibilities.

Due to their expertise, however, a strong connection exists between engineers/ engineering activities and the potential for resulting physical harm; although, even here, no absolute standard is possible. The claim that engineers have a responsibility for the physical safety of the public requires clarification.

Again, despite being a term often taken to have a clear meaning, "safety" is a rather complicated phenomenon. Safety can be understood in terms of its subjective and objective dimensions. Subjective safety consists in the *feeling* of not being in danger. Objective safety consists in the *fact* of not being in danger. Although one can *feel* perfectly safe, no one can ever *be* perfectly safe.

From an engineering perspective, the first thing to realize is that no product can ever be made perfectly safe—in an objective sense—and be economically viable. Second, not all possible consequences can be foreseen. This means there will always be *degrees* of safety associated with engineering activities. Objectively, safety is a concept that operates on a sliding scale: things can be more or less safe, so being objectively safe does not have any one, clearly defined state. Third, for members of the public, the feeling of safety is influenced by knowledge, since a feeling of subjective safety may not correspond to a high degree of objective safety or vice versa. Although many are afraid of and feel unsafe flying in airplanes, for example, statistically speaking and objectively, air travel is one of the safest forms of transportation. As responsible experimenters, engineers should consider both the objective and subjective dimensions of safety.

Objectively, safety is a matter of "risk," which can be defined as the potential that something unwanted and harmful could occur—the likelihood of failure multiplied by the severity of the consequences of failure. As with harms, there are different types of risks: bodily, psychological, economic, environmental, and so on. Again, as responsible experimenters, engineers should be concerned with all of these. However, there are clearly limits to engineers' abilities to assess the extent of all risks or even to discover them. Imagine, for example, that you are designing a new four-wheel drive vehicle—a model of the Ford Explorer, for instance—and try to think about all the risk factors you would have to consider in your design. To carry out a risk-benefit analysis, you would need to consider the following:

1. What kinds of damages are possible?
2. What are the potential severities of these damages?
3. What are the probabilities that human beings would be exposed to these damages?
4. What are the technical feasibilities of the alternatives?
5. What are the economic feasibilities of these alternatives?
6. What are the potential adverse effects of these alternatives?[79]

---

79. For an extended treatment of risk analysis in public engineering projects, see Thompson and Perry (1992), Ayyub (2014), and Millar and Lessard (2001). Treatments of considerations of risk also vary within different domains of engineering. For an example concerning technological design, see Star (1969). For an example from civil engineering, see Faber and Stewart (2003).

Further, while safety is a central feature of the design process, engineers should also be concerned with other factors that influence design, such as efficiency, costs, longevity, and manufacturability. All of these factors form parts of larger cost-benefit analyses, only some aspects of which are directly related to engineering responsibilities. Especially since—to a large extent—they deal with the future, analyses of risks and costs are never completed once and for all. In terms of risks, however, engineers are in the best positions to make such judgments, and doing so is part of their professional responsibilities. Making these judgments concerning risks requires a broad basis of knowledge on the parts of engineers—the study of technical materials alone is insufficient. For this reason, ABET requires that accredited engineering programs and those seeking accreditation can demonstrate that that their students possess "the broad education necessary to understand the impact of engineering solutions in a global, economic, environmental, and societal context" (Criteria 2016–17).

Like safety, assessments of risks have subjective components. These components can be framed in terms of the acceptability of risks—the level of risks resulting from the activities of engineering, which the public deems acceptable. The social and cultural values of particular societies play a major role in levels of risk acceptance. For example, questions regarding the value of individual human lives, importance of future generations, how risks should be distributed among the population, comparisons of different risks, circumstances in which risks will occur, etc., all determine the acceptability of risks. The following outlines general criteria of risk assessment within the Western world:

Higher acceptance of risks	Lower acceptance of risks
Voluntary	Involuntary
In control	Not in control
Occupational	Nonoccupational
Common hazard	Dread hazard
Effect later	Effect immediate
Risk known	Risk unknown
Consequences reversible	Consequences irreversible
Statistical risk	Known individual at risk[a]

[a]*Regarding criteria of risk assessment along these lines, see, for example, Renn (1998). Concerning the ways these differ cross culturally, see Renn and Rohrmann (2000).*

Thus, engineers should be concerned with not only the risks but also the acceptability of risks. Given that levels of risk acceptance vary culturally—in ways over which engineers do not have control—to a reasonable extent, they should be aware of cultural conditions in fulfilling their safety obligations. Obviously, however, there are limits beyond which engineers—especially on an individual basis—cannot know all the objective risks associated with courses of actions or the acceptability of these risks. How then can engineers possibly be responsible for the consequences associated with engineering activities?

One possible solution is "informed consent," a doctrine generally accepted within US medical contexts and incorporated into laws. Informed consent requires that physicians inform patients of the known risks of procedures, and that—as reasonable beings—patients understand these risks and voluntarily consent to the procedures. Without informed consent, the procedures will be considered "battery," unlawful touching. In dealing with questions of safety, it has been suggested that engineering adopt a similar model: without its consent, the public has a right not to be harmed. Therefore, engineers would have the obligation of informing the public of possible harms that would result from engineering activities. As physicians generally deal with patients on a one-on-one basis in medical contexts, obtaining this consent is easier. By contrast, in engineering, products or processes are generally introduced on a society-wide basis, where the end users or those affected are not well known.

First, engineers thus seem morally and professionally required to ensure that their designs and products are the safest possible, within the constraints of risk-benefit analyses. This requires the adequate testing of products and processes before being marketed, and perhaps even initial test marketing on small or localized scales. Second, engineers are responsible for informing the public of known risks associated with engineering activities and where the risks associated with these activities cannot be determined. Adequate warning labels and instructions for the use of products and processes are, therefore, required. Third, as experimenters, engineers should be aware of changes in the uses of products and processes they create, as well as changes in the environments in which these take place. This could entail, for example, making changes to the products or processes, or to their instructions. Assuming these conditions are met, members of the public can make voluntary decisions regarding whether or not to use the products or processes engineers have created, formulating their own decisions concerning the acceptability of the risks involved:

- Could there be discrepancies in measurements of safety between cultures? If so, explain how this could affect global engineering and how it might be overcome.
- Levels of risk acceptance often vary from culture to culture. How should engineers measure the acceptability of risks in general? What other factors should be considered when measuring the acceptability of risks in particular countries?
- It is relatively easier for medical practitioners to inform and obtain the consent of their patients. How might larger, engineering organizations inform society?

5.3 THE CONNECTION OF SAFETY WITH OTHER RESPONSIBILITIES

Public safety is generally—although not universally—recognized by the global community as a core ethical responsibility of engineers. Engineers who fail to consider public safety in their engineering activities are incompetent engineers. However, aside from incompetence, engineers might neglect considerations of safety because other factors are allowed to influence their decisions—factors external to the realm of engineering. As explained above, evaluating subjective dimensions of safety can present engineers with potential difficulties, and the expressed desires and wants of the public are important to this evaluative process. Do engineers have the responsibility of preventing the public from fulfilling its desires—thereby acting paternalistically—if they believe the fulfillment of these desires would create an objectively unsafe situation? Whose interpretation of safety should count in this decision-making process, and to what extent?

More significantly, engineers have responsibilities to others, which are not based solely on engineering. As was mentioned in earlier chapters, each human being plays many roles. Two additional, major roles that engineers often play are those of employee and family member. These other primary roles can create "conflicts of interests," situations where the legitimate demands of two or more roles conflict, so that not all of one's duties can be met. Traditional engineering ethics has held that engineering duties—especially that of public safety—should always take priority. Given the implications and consequences of failing to fulfill the duties associated with other roles, however, this seems like a potentially unreasonable expectation.

Traditional ethical theories typically hold that any one life is equal to that of any other. Hence, if an engineer could save the lives of *two* persons by fulfilling his or her engineering duty, for instance, versus saving the life of *one* of his or her children, but neglecting the engineering duty, then ethical theories would typically demand that the engineer fulfill his or her engineering duty and save the two lives. However, from the perspective of role responsibilities—discussed before—the situation is less clear, since the duty of a parent is, first and foremost, for the welfare of one's own child or children. Were there only one child, the parental role would be completely destroyed with the death of the child. In this type of situation, for most parents, the choice would be obvious.

Should one be willing to condemn a parent in a situation such as this for acting unethically? If not, then it is necessary to recognize that engineering duties cannot be understood in isolation, simply condemning any action that runs contrary to engineering duties as an illegitimate conflict of interests. This becomes more pressing given the wider contexts of typical engineering practices, since the vast majority of engineers are also employees. For these reasons, it is necessary to examine the roles of engineers in business environments, a task carried out in the next chapter:

- State and explain an actual incident of which you know where engineering or technology companies neglected safety considerations to meet public desires. In this case, what principles of ethical global engineering were violated?
- As an engineer working on the maintenance and safety of the levee system in Louisiana, would it have been ethically correct to warn his or her family and friends but not the general public? As an engineer—versus a friend or family member—which ethical responsibilities might come into conflict? Explain your answers.

EXERCISE TWO — HOW SAFE IS SAFE? (PART TWO)

Returning to the case of Hurricane Katrina, complete steps 3–10 of the case-study procedure on the following sets of ethical issues:

- If engineers believed the design constraints of the levee system to have been inadequate, then should they have built them according to these constraints? To what extent do engineers have a responsibility to ensure the safety of the local population, despite whatever countervailing demands might exist?
- Should the Corps of Engineers have been involved in evacuation planning rather than its having been left to local government agencies? What are the responsibilities of engineers to ensure that people have safe exits if disasters occur in relation to engineered work?
- Was the local population sufficiently aware of the potential risks associated with the levee system? Should engineers think of themselves as "social experimenters," with a corresponding duty to gain the informed consent of populations affected by their actions?
- Even though control of the levee system had been turned over to various local agencies, did the Corps of Engineers have a responsibility for continued follow-up? What is the responsibility of engineers to monitor the ongoing status of completed projects they have designed?
- Should early developers of the levee system have forecast the growth of the New Orleans area and the attendant problems this would create? To what extent can engineers be expected to factor in the long-term consequences of their decisions, even though many of these cannot be known?
- Are the levee designers responsible for factoring in the effects of dredging—removing sediment from the bottom of—the Mississippi and the loss of wetlands? To what extent can engineers be expected to be aware of the interactive nature of such decisions?
- Did the engineers who designed the levees have a responsibility to use the best available model—from the Netherlands—in their decisions? What duties do engineers have to be aware of, and utilize, prior knowledge in their designs?

CASE STUDY TWO—THE UBER RAPE SCANDAL: USER SAFETY AND THE RESPONSIBILITIES OF TECHNOLOGY FIRMS IN GLOBAL CONTEXTS

On Dec. 6, 2014, a 35-year-old female financial analyst working in New Delhi, India, accused Shiv Kumar Yadav, a 32-year-old male Uber driver, of raping her. On Dec. 9, 2014, the Indian government announced a ban on Uber's services in New Delhi—shortly thereafter, governments in Spain, Thailand, and the Netherlands followed suit—and in Jan. 2015, Uber was sued in a US court for "failing to ensure" passenger safety ("India Woman," 2015).

The India case raises questions regarding user safety in the age of digital apps and the "sharing economy," as well as the responsibilities of technology firms in these increasingly decentralized and global contexts—especially given national legal differences.

"According to the victim, she had hailed [a] cab using Uber's mobile-phone-based application about 9:30 p.m. on Dec. 5" (Koutsoukis, 2014). She fell asleep in transit, awaking to find the car stopped and Yadav assaulting her. She "tried to escape," but "the driver threatened to kill her and then raped her" (Koutsoukis, 2014). For Yadav, this was not an isolated incident.

Approximately a month before, in Nov. 2014, Nidhi Shah reported Yadav to Uber. Shah complained that Yadav "kept staring at Nidhi and her partner through the mirror," which made them uncomfortable (Koutsoukis, 2014). Uber sent an email to Nidhi, promising to follow up the complaint and give feedback to Yadav. Those familiar with him were unsurprised.

Yadav was described as a "compulsive sex offender" (Koutsoukis, 2014): "you won't find a single household" in the village in which Yadav was born and raised "whose women he hasn't teased or molested" (Koutsoukis, 2014). The villagers rarely made "any criminal complaint because they" believed it would "bring a bad name to the village" (Koutsoukis, 2014)—despite molestation allegations dating back as far as 2003 (Singh & Jethro, 2014).

Yadav was "previously accused in three more criminal cases," two of rape and one of molestation ("Uber Rape," 2014). In 2011, he "was acquitted of charges of raping a woman at knifepoint due to an apparent lack of evidence," and in 2013, "he was granted bail after another rape case was registered against him" (Koutsoukis, 2014). Despite this background, Yadav was hired as a driver at Uber. During his time as a driver, reportedly, Yadav "was always on the prowl for women traveling late at night, especially those who he thought were vulnerable" ("Uber Rape," 2014).

"Unlike other more traditional taxi services that require drivers to undergo police checks before granting them a taxi permit," Uber in India did not screen its drivers (Koutsoukis, 2014)—despite Uber management in India having claimed that all of its "driver partners are put through a rigorous quality control process" (Koutsoukis, 2014).

After this incident, officials in the Indian government said "top executives" at Uber "could face charges of criminal negligence over the rape for not conducting adequate background checks of its drivers," although this never came to fruition (Koutsoukis, 2014). On Feb. 2, 2015, in the aftermath, Uber promised to "establish clear background checks currently absent in their commercial transportation licensing programs" (Russell, 2015), and Yadav's victim voluntarily dropped her US-based lawsuit against Uber. Although Uber was not ultimately found guilty in either criminal or civil proceedings, it undoubtedly suffered in the court of public opinion, bringing bad PR to the company.

This incident emphasizes the complex nature of safety within engineering, raises questions regarding the ethical responsibilities technology companies have to their users, and highlights important differences between legality and ethicality:

- As a result of traffic accidents and robberies—through user inattention/distraction and thieves luring players to secluded areas—app-based, augmented reality games such as PokémonGo have raised questions about the responsibilities of app developers and technology firms to user safety. How are these concerns similar to and different from those regarding apps such as Uber and services such as Airbnb? Explain your answers.
- Aside from the ethical ramifications, describe how the rape scandal in India might affect Uber's "bottom line," the profits a company makes from the sales of its goods/services. Why?

EXERCISE THREE—THE UBER RAPE SCANDAL

Based on principles and contents discussed so far, complete steps 1–10 of the case-study procedure on the Uber Rape Scandal.

5.4 SUMMARY

Aside from simply identifying individuals or organizations responsible—and on which to lay blame—for engineering-related disasters, studying cases such as these allows one to learn about what went wrong, and how to better prevent disasters from occurring in the future. The case of Hurricane Katrina demonstrates the huge numbers of individuals and organizations involved in large-scale engineering projects and those adversely affected if failures occur. As was discussed in Chapter 4, a number of specific engineering ethical duties follow from the primacy of safety, although safety can itself be an opaque notion. To better understand the nature of safety and its relation to engineering ethical duties, the analogy of engineering to social experimentation highlights the need for—but difficulties involved in—obtaining the informed consent of those affected. Especially in cross-cultural environments, matters become more complicated when considering the fact that safety involves not only objective but also subjective dimensions. Although public safety is generally recognized as a core engineering duty, as human beings, engineers occupy various roles, and, at times, the duties associated with these other roles can conflict with those of engineers. As the case of Uber illustrates, engineering and technology firms have a responsibility to ensure the safety of their users, although the extent of this responsibility and the manner in which they should do so might not always be obvious.

REVIEW QUESTIONS

1. How could one argue that, as a disaster, Hurricane Katrina falls outside the scope of engineering ethics?
2. In what manner did the Corps of Engineers lack autonomy in making decisions regarding the levee system?
3. With reference to the three main causes of the Hurricane Katrina disaster in New Orleans as identified by the Independent Levee Investigation Team, what basic ethical principles of global engineering were violated? Explain your answer.
4. When learning about ethics in engineering, what are two pros and two cons associated with studying large-scale disasters?
5. List and explain three characteristics of engineering in terms of which it could be understood as social experimentation. What responsibilities stem from such an analogy?
6. Explain the difference between objective and subjective safety.
7. What are three conditions engineers should meet to ensure members of the public can make voluntary decisions regarding the designs and products of engineers?
8. Explain one action engineers can take to become responsible for the objective risks associated with their products and/or designs.
9. Give some examples of how particular social and cultural values play roles in levels of risk acceptance.
10. With regard to the second case study, explain how Uber could have been considered negligent or irresponsible.
11. Describe how the case of Uber in India highlights the difference between legality and ethicality, and the need to consider ethics in addition to the law.

REFERENCES

Ayyub, B. (2014). *Risk analysis in engineering and economics*. Boca Raton, FL: CRC Press.
Criteria for Accrediting Engineering Programs. (2016–2017). ABET. http://www.abet.org/accreditation/accreditation-criteria/criteria-for-accrediting-engineering-programs-2016-2017/.
Faber, M., & Stewart, M. (2003). Risk assessment for civil engineering facilities: Critical overview and discussion. *Reliability Engineering & System Safety, 80*(2), 173–184.
Graumann, A., et al. (2005). *Hurricane Katrina: A climatological perspective*. NOAA National Climatic Center. http://www.ncdc.noaa.gov/oa/reports/tech-report-200501z.pdf. October.
India Woman Sues Uber over Driver Rape Allegation. (2015). *BBC News*. 30 January. http://www.bbc.com/news/world-asia-india-31052849.

Koutsoukis, J. (2014). Uber rape scandal in India grows. *Brisbane Times*, 12 December. http://www. brisbanetimes.com.au/world/uber-rape-scandal-in-india-grows-20141212-125ppz.html.

Luegenbiehl, H. (2007). Disasters as object lessons in ethics: Hurricane Katrina. *IEEE Technology and Society Magazine, 26*(4), 10–15. http://ieeexplore.ieee.org/document/4408563/?part=1 Winter.

Martin, M., & Schinzinger, R. (2010). *Introduction to engineering ethics*. New York, NY: Mc-Graw Hill.

Millar, R., & Lessard, D. (2001). Understanding and managing risks in large engineering projects. *International Journal of Project Management, 19*(8), 437–443.

PBS (2006). NewsHour. 12 June http://www.pbs.org/newshour/bb/science/july-dec05/levees_10-20.

Petroski, H. (1985). *To engineer is human*. New York, NY: St. Martin's Press.

Renn, O. (1998). The role of risk perception for risk management. *Reliability Engineering & System Safety, 59*(1), 49–62.

Renn, O., & Rohrmann, B. (2000). *Cross-cultural risk perception: A survey of empirical studies*. (Vol. 13). New York: Springer Science and Business Media.

Russell, J. (2015). Uber is finally applying more stringent driver background checks in India. *Tech-Crunch News*, 2 February http://techcrunch.com/2015/02/02/uber-is-finally-applying-more-stringent-driver-background-checks-in-india/?ncid=rss.

Seed, R., et al. (2005). *Investigation of the performance of the New Orleans flood protection systems in hurricane Katrina*Draft Final Report, No. UCB/CCRM-06/01. 29 August. http://www. ce.berkeley.edu/~new_orleans/report.

Select Bipartisan Committee to Investigate the Preparation for and Response to Hurricane Katrina. (2006). *A Failure of Initiative*. 15 February. http://katrina.house.gov.

Singh, H. & Jethro, M. (2014). Police: Indian uber driver accused of rape is awaiting trial in other cases. *CNN*. 11 December http://edition.cnn.com/2014/12/10/world/asia/india-uber-rape-case/.

Star, C. (1969). Social benefit versus technological risk. In T. Glickman & M. Gough (Eds.), *Readings in risk* (pp. 183–194). New York: Johns Hopkins University Press.

Thompson, P., & Perry, J. (1992). *Engineering construction risks: A guide to project risk analysis and assessment implications for project clients and project managers*. London: Thomas Telford.

Uber Rape: Driver Involved in Previous Rape Case. (2014). *SifyNews*. 9 December. http://www.sify. com/news/uber-rape-driver-involved-in-previous-rape-case-news-national-omjvu0ifejgga.html.

US Army Corps of Engineers (2006). Performance evaluation of the New Orleans and Southeast Louisiana hurricane protection system—Final report of the interagency performance evaluation task force. Volume I—Executive summary and overview. 1 June. http://biotech.law.lsu.edu/katrina/ipet/Volume%20I%20FINAL%2023Jun09%20mh.pdf.

Chapter 6

The Global Business Environment: What Engineers Should Know

Chapter Objectives

Having read this chapter, completed the included exercises, and answered the associated questions, readers should be able to

- with reference to the case of Toshiba Machine Tools, note and explain conflicts in values that can arise when conducting business in global environments and the importance of perceived differences in cultural and social values in these contexts;
- describe the role of and justification for the importance of ethics in business contexts and how these would be different from engineering ethics in general;
- with regard to business environments, list and explain the justifications for the six principles of ethics for organizations and the six principles of ethics for employees, giving examples of where they were or were not followed in the Toshiba Machine Tools case, and why they would be important;
- explain how engineering functions within the context of business environments, noting points of convergence and divergence in business and engineering ethics, as well as potential conflicts between the two;
- describe the importance of ethical engineering to business, noting its relation to brand recognition and national reputation, with regard to the case of the Volkswagen emission scandal.

CASE STUDY ONE—WHEN BUSINESS AND POLITICS COLLIDE: THE CASE OF TOSHIBA MACHINE TOOLS

In 1987, one of the most prominent pictures of the year was US House of Representative members smashing a Toshiba boom box on the steps of the Capitol in Washington, DC. This demonstration was, in part, a display of frustration with long-running US-Japan trade disputes, although its immediate cause was the involvement of a Toshiba subsidiary in the sale of technology to the Soviet Union. As Chapters 4 and 5 explain, certain engineering ethical responsibilities follow from the primacy of public safety. However, contemporary engineering activities rarely take place in isolation. They almost always occur in broader contexts, where engineers occupy multiple roles and have to navigate competing and—at times—conflicting duties that result from these roles. As with Hurricane Katrina, the Toshiba Machine Tools case continues to introduce readers to these broader contexts, specifically, those of business and politics.

The case of Toshiba Machine Tools had its origins in the work of the now infamous John Walker spy ring. In 1980, Walker alerted Soviets to the fact that the major reason it was easy for the United States to detect Soviet submarines was the noise their propellers made. US listening posts were able to detect the submarines up to 200 mi away, with 90% of the noise produced by the propellers. As a result of the information obtained from Walker, the Soviets explored ways of lowering sounds from the propulsion systems. The solution lay in more accurate and standardized milling of the propellers—grinding the propellers into shape. At the time, this was beyond the technical capabilities of the Soviet tool industry, but not that of the Japanese (Triplett, 1988, pp. 9–10).

After initial contact with Wako Koeki, a Japanese trading firm, on Apr. 24, 1981, a contract was signed between the following parties: C. Itoh & Co., Japan's largest trading firm; Toshiba Machine, 50.2% owned by the Toshiba Corporation; Kongsberg Vaapenfabrikk, a state-owned subsidiary of the Norwegian Ministry of Industries; the Soviet Technological Machinery Corporation; and a number of KGB agents (Triplett, 1988, p. 8). The contract called for the delivery of eight computer-controlled milling machines, four 9-axis machines, and four 5-axis machines. Toshiba Machine was to provide the hardware and Kongsberg the controllers for the 9-axis machines, with Toshiba Machine supplying its own controllers for the others. The total value of the contract was $17 million (Cook, 1987, p. 42; Kapstein, 1987, p. 65).

Although this appeared to be an ordinary business arrangement, it was illegal under both Japanese and Norwegian law. Both countries were signatories to the Coordinating Committee for Multilateral Export Controls (CoCom), which was established to control the sales of strategically important western technology to the Eastern-Bloc countries—countries allied in various ways with the Soviet Union. The membership of CoCom consisted of all of the NATO countries, with the exception of Iceland, and the addition of Japan. Under CoCom regulations, it was illegal to export any milling machines with more than two axes to Soviet-bloc countries (Goldman, 1987, p. 20).

The 9-axis machines were huge, standing three stories tall and weighing approximately 250 tons each. They were capable of grinding in nine different directions at the same time, with a tolerance of less than 0.01 mm. The machines could do so on propellers of up to 40 ft in diameter. They had a special feature that allowed them to grind both faces of a propeller blade simultaneously, thus decreasing the deformation caused by pressure from the grinding process and producing a much thinner blade than would otherwise be the case (Triplett, 1988, p. 9).

At the time, all machine tool exports from Japan to the Soviet Union required an export permit. For this reason, instead of permits for the 9- and 5-axis machines actually shipped in 1983 and 1984, C. Itoh & Co. applied for a permit to send a set of TDP 70/110 model two-axis machines, thereby circumventing CoCom regulations (Goldman, 1987, p. 20). Kongsberg also falsified its export papers. The claimed destination for the machines was an electric power plant in Leningrad, but to "install and demonstrate the machinery, Toshiba and Konesberg (sic) technicians traveled not to an electric power plant, but to Leningrad's highly secret Baltic shipyards" (Goldman, 1987, p. 20). As a result of the sales, Soviet submarines reduced their noise-detection levels from the previous 200 to 10 mi (Triplett, 1988, p. 10).

In Dec. 1985, details of the sale began to emerge when Hitori Kumagai, a "disgruntled" employee and Moscow manager of Wako Koeki at the time of the sale, wrote a "whistle-blowing" letter to the chairperson of CoCom (Koepp, 1987, p. 53). "Kumagai named names and dates, identified equipment by model number and destination, and described the intended Soviet use. He signed the letter including his address and even his telephone number. His letter included copies of all the secret contracts and an inch-thick technical attachment of engineering drawings" (Triplett, 1988, p. 10). CoCom forwarded the information to the Japanese Ministry of International Trade and Industry (MITI), an agency responsible for export controls. It replied that Kumagai's charges were groundless (Triplett, 1988, p. 10). In 1986 and 1987, the Japanese MITI continued to deny the charges after being approached on a number of occasions by US officials.

On Apr. 28, 1987, *The Detroit News* published an investigative report based on leaked information, finally bringing the case to public attention. Within days, police raided sites of the Japanese companies involved. Initially, the president of Toshiba Machine denied the charges. Later, it was reported that, inside "the company, a full-scale cover-up was under way, in which employees incinerated documents by tossing them into factory furnaces" (Koepp, 1987, p. 53). By the end of May, two executives from Toshiba Machine had been arrested, and on Jul. 1, Sugiichiro Watari, the president of Toshiba Corporation, and Shoichi Saba, its chairman, both resigned ("Hard Pounding," 1987, p. 64).

These rather drastic actions by Toshiba Corporation executives were, in large part, the result of US reactions to publicity surrounding the case. Just prior to these resignations, the US Senate passed a bill—with a 92–5 vote—preventing the Toshiba Corporation and Kongsberg from selling any products in the United States for 2–5 years. Estimates placed Toshiba losses in US sales at $2.3 billion per year, and losses in Toshiba US operation jobs at 4000 ("Making Amends," 1987, p. 49). Estimates placed costs associated with the United States regaining the technological edge it had lost to the Soviet Union due to the sales of these technologies at $8–40 billion (Triplett, 1988, p. 8; "Toshiba Bans," 1987, p. 27; Copeland, 1987, p. 40).

Toshiba Corporation responded to the Senate bill with a $100 million advertising campaign in the United States, including full-page apologies in

60 newspapers, with the headline *"Toshiba Corporation Extends its Deepest Regrets to the American People"* (Chandler, 1987, p. 11). It also rallied the support of Toshiba-related suppliers and customers in the United States and the support of officials in states with US plants (Dryden, 1987, p. 58). In response to the crisis, the Norwegian government broke up Kongsberg, selling off its assets to 13 other companies ("Norway Is," 1987, p. 32). For its part, the Japanese government promised to tighten its export control system, adding new inspectors, extending the statute of limitations on export violations, and punishing the Japanese companies involved (Triplett, 1988, p. 12). By the fall of 1987, in the United States, as a consequence of these actions, the possibility of sanctions and retaliation faded from public consciousness (Dryden, 1987, p. 58). In Japan, however, the effects were longer lasting, demonstrating a different perspective on the case: "The sledgehammer scene, which was largely ignored by the American media, was shown over and over again in Japan, to the point where it now lodges uneasily in the collective national consciousness" (Packard, 1987, p. 348).

While many in the United States viewed the sale of the milling machines as an issue of international security and another instance of Japanese companies putting profit above all else, most in Japan saw it as another in a long series of occasions by trade competitors to bash Japan. To the Japanese, the US response ignored widespread violations of export regulations by countries other than Japan. Even in this particular case, the role of the Norwegians seemed to be largely ignored. A US writer reflected this sentiment: "The Soviet Union and Norway played key roles in the intrigue, too. But no one in Washington suggested banning wheat sales to Russia—and no one was smashing cans of Norwegian sardines outside the Capitol" (Copeland, 1987, p. 40).

The Japanese were particularly upset that Toshiba Corporation was being held responsible for the actions of an only partially owned affiliate: "It's terrifying to think that we could be held responsible for an affiliate's wrongdoings" (Armstrong et al., 1987, p. 86). Indeed, many of the articles referred to here abbreviate "Toshiba Machine" as "Toshiba," such that confusion among the US public and political establishment could be expected. Perhaps, most crucially, however, was the fact that Toshiba Corporation's top two officials resigned as an act of responsibility for the illegal sale. "In American terms, the resignations of the top two executives amount to an admission of guilt on the part of Toshiba. But it is difficult to agree that the responsibility extends to the parent company, Toshiba" ("Toshiba vs.," 1987, p. 18). In taking the typical Japanese action of resigning as an apology for the entire organization, Japanese commentators felt that Toshiba had opened the way for Americans to vent their anger over trade relations: "Meanwhile, back in Japan, commentators were suggesting—with almost no dissenting opinion—that the United States had made Toshiba a scapegoat for its own economic problems" (Chandler, 1987, p. 11).

Economic issues certainly played a major role in the case. Even while the US Department of Defense argued for making an example of Toshiba, the US

Department of Commerce was concerned about the economic effects of potential sanctions. These concerns eventually won out (Dryden, 1987, p. 58). This undermined the arguments of those who sought to blame the Japanese government for the entire matter, focusing on the role of the MITI in trade transactions. They alleged that the MITI not only stalled any serious investigation of the sale but also advised Toshiba Machine on how to circumvent CoCom restrictions, according to *Yomiuri Shimbun*, a leading Japanese newspaper (Chandler, 1987, p. 12). This case also initiated a wider discussion regarding the role of international trade restrictions on transfers of technologies. As one commentator mused, "technology now moves so fast that what is exotic and strategically important today is likely to be mundane and easily accessible tomorrow" (Goldman, 1987, p. 73):

- Discuss the function of apologies in different cultures. What lessons about the internationalization of engineering can we learn from this?
- Do you believe Toshiba Corporation shared in the responsibility for what happened with Toshiba Machine? Why or why not?

EXERCISE ONE—THE CASE OF TOSHIBA MACHINE TOOLS (PART ONE)

Complete the case-study procedure with regard to information given in the case of Toshiba Machine Tools above.

6.1 ENGINEERING AND BUSINESS ENVIRONMENTS

This chapter examines business environments. Since a great deal of engineering activities occur in such environments, it is important for engineers to take into account principles of business ethics in making ethical engineering decisions. Again, this reflects the somewhat atypical approach taken here: considering engineering ethics in a contextual fashion. Acting as though engineering occurs in isolation—apart from other aspects of social and economic life— is unreasonable. In this regard, the most important and influential factor to consider is the nature of the working lives of engineers, most of whom are employees of companies. Before being able to discuss engineering ethics in business contexts, however, clarification regarding the nature of business is necessary.

6.2 THE NATURE OF BUSINESS

Business exists to produce and develop goods and services for exchange, exchanging these for other goods, services, and money. Socially determined criteria form the specific basis in terms of which business is organized—in other words, business need not exist in any one form. Questions concerning the forms business should take to maximally benefit human existence have given rise to historical conflicts. Thus, to further explore business environments in a general sense—as with a global approach to engineering ethics—several assumptions are necessary.

From a theoretical and prescriptive perspective, significant debates exist regarding the ethics of different types of business practices. From a practical and descriptive perspective, however, worldwide business currently exists in one dominant form: private enterprise aimed at profits. This is the first assumption, as most global business occurs or is moving in this direction. This includes both China and India, the world's two most populous countries, which previously emphasized state-owned and state-controlled business institutions. Although considering business in terms of privately owned and capitalistically motivated enterprise is by no means a necessary assumption, given its relevance from a global perspective, this is a useful assumption.

Even within the framework of capitalism, however, a further distinction is necessary: businesses where the owners also run the company and enterprises where ownership is separate from management. The latter generally relies on investors who hope to profit, having taken on financial risk by investing and potentially losing their money. This is especially true of public companies, which raise money by issuing and selling stocks in the company through initial public offerings (IPOs).

Especially early in their careers, engineers typically work in enterprises run by management in large-scale corporations. Thus, the second assumption is as follows: ethical issues related to employees and managers in such corporations are the most appropriate focus. Making this assumption regarding the form of business to use as a foundation for analysis does not, however, imply a judgment about this form being right or any other form being wrong. This decision simply reflects the current state of affairs and anticipated trends in the near future.

The final necessary assumption is as follows: an individual corporation is itself a legal entity apart from its employees. In US law, for example, corporations have the status of "persons." Corporate managers thus act as agents for this entity and are bound to represent its interests, even though, in reality, corporations are owned by thousands, if not millions, of individuals. This fact often raises questions regarding in whose interests a business is actually being run, those of investors or managers.

Although this may be in dispute, on the "agency theory" businesses should be run in the interests of investors.[80] Given the assumption about human nature from Chapter 1, these investors are primarily interested in gain for themselves in return for their investments. Investors are interested in profits. The question then arises concerning the extent to which ethical considerations should limit the pursuit of profits.

---

80. Regarding agency theory and the nature and history of corporations, see, for example, Shapiro (2005), Eisenhardt (1989), Davis (2013), and Chandler (1990).

6.3 THE ROLE OF ETHICS IN BUSINESS

Some have argued that ethics does not apply in the realm of business, perhaps most prominently Albert Carr. They claim that business is like a game with its own separate set of rules, the laws according to which businesses operate. An analogy to this situation would be the following: in normal life, one person punching another for no reason would be unethical. In the boxing ring, by contrast, that would be expected and not unethical.[81] In a similar vein, the analogy runs, although taking advantage of someone is generally considered unethical, in business, this would be expected, assuming no laws are broken.[82]

The claim that business can be separated from other aspects of life, however, seems suspect. In business, the same kinds of ethical rules are needed as in life in general. Otherwise, institutions fail to function. Physically harming a competitor to eliminate competition is just as wrong in business as physically harming a rival for a love interest—aside from any laws regulating behaviors. Furthermore, the effects of business activities are not limited to the realm of business. Business activities clearly have serious consequences in all other aspects of life. Business actions thus have ethical import. To an individual harmed, the source is irrelevant; the harm still exists. Finally, when looking at the world, people obviously make ethical judgments about the actions of people in business, a trend gaining in momentum.[83]

In business contexts, another question concerns to what or whom ethical responsibilities should be assigned. If corporations are persons, then they are persons in rather unusual ways, specifically, ones that only act through other actual persons. Thus, should corporations be held accountable for their actions? Should managers be held responsible—since they act on behalf of corporations but are not actually corporations themselves?

In terms of ethical responsibilities, answers to these questions have clear implications for determining and assigning degrees of praise and blame. If only corporations should be held responsible, then this undermines moral justifications for laws punishing corporate managers and other forms of ethical judgments. Additionally, on this view, corporate managers should only be responsible for actions that run counter to the wills of corporations. This result seems unsatisfactory, however, since actions are ultimately only ever decided by actual persons. For these reasons, ethical responsibilities should govern the actions of managers, even when they are acting on behalf of corporations.

A final question concerns toward whom or what managers have ethical responsibilities. A common form of framing this debate is the "shareholders" versus "stakeholders" perspectives. The shareholders view asserts that managers are only responsible to the interests of investors—to seek profits as long as

---

81. See especially Carr (1968) concerning this view.
82. On this perspective, see Hasnas (1998).
83. This has been especially true since the 2007–09 financial crisis and Great Recession.

ethical rules are maintained. The stakeholders view asserts that managers are ethically accountable to all of the stakeholders of corporations—all those affected by the actions of corporations. Typical stakeholders would include employees, suppliers, customers, governments, local communities, and so on. Against the traditional dominance of the stockholders view, the stakeholders perspective is becoming ever more prevalent.[84] This view is clearly more in line with the claim that ethics is about actions that have the potential to seriously impact the lives of others—as opposed to simply considering the effects of actions on a few, as does the stockholders view. If one grants that ethics should have a role in business, then balancing the effects of business actions on various constituencies should be considered:

- Can you think of actions or practices that would be ethical in business environments but unethical outside of business environments? If so, then list and explain these.
- Why do you think the role of a manager is important when considering stakeholders and stockholders?

6.4 ETHICAL PRINCIPLES FOR BUSINESS: THEIR DIFFERENCES FROM ENGINEERING ETHICS

Once the claim is accepted that ethics should have a role in business, the next step consists in determining the nature of ethical actions in this context. Developing a few basic principles for ethical business is possible, although this development is not as complete as it should be in relation to engineering. These principles will be established on a foundation different from that of engineering ethics: a combination of role responsibilities and the broad ethical duties that follow from being human in general, rather than an engineer specifically.

This difference is based on the fact that expertise is typically not required to engage in business—with the exception of certain business functions, such as accounting—unlike the expertise required to be an engineer. To run a business well, it might be argued that special training is helpful, for example, completing an MBA degree. However, expertise is clearly not a requirement, as can be seen in counterexamples: Bill Gates and Mark Zuckerberg both amassed fortunes after dropping out of university. By contrast, education in engineering is a minimal requirement to be an engineer. Business ethics is, thus, a form of more traditional applied ethics, where general ethical principles are applied to specific contexts, rather than one where special responsibilities follow from the specialized nature of business itself.

As was mentioned above, businesses can be considered in two ways: from the perspective of organizations and from the perspective of individuals. In the case of the second, a further distinction is necessary: managers—those who

---

84. For an overview of these two positions, see Smith (2003).

make decisions on behalf of the organization—and other types of employees who follow directives. Managers are simultaneously both agents and employees of corporations.[85] The applications of these ethical principles are different for corporations and employees. The following examination takes place along the lines of this division, in terms of organizational ethical principles and ethical principles for employees.

The ultimate aim of profits within business does not eliminate general ethical obligations. Those ethical principles that apply in everyday life should also apply in the realm of business, with the addition of special role-related obligations: for managers, the primary role is based on the relation between superior and subordinate; a position of authority creates special obligations on behalf of the organization. Likewise, a position of subordination creates special—although different—duties for employees. Thus, the following is a list of organizational ethical principles that seem important to highlight in the contexts of business, along with justifications for their importance.

6.5 ORGANIZATIONAL ETHICAL PRINCIPLES

The following list of ethical principles for organizations is not the same as those generally included in the now common "value statements" of organizations. These value statements embody the ideals toward which the organizations and their members should strive—what particular organizations stand for. These value statements embody attempts to go beyond the minimal level of ethics expected from all corporations in general; in other words, they represent superior sets of values.

6.5.1 Corporations and Harms: Through Their Actions, Corporations Should Endeavor to Avoid Producing Unnecessary Harms to Those in and Outside of Their Organizations

The basis for this principle is the more general claim that human life should be respected, discussed in previous chapters. As was noted before, however, the interdiction against "harm" is not absolute, especially the potential for financial harms resulting from business operations. If such harms were to be avoided altogether, then businesses would cease to operate. Thus, it is necessary to balance the potentially positive and negative effects that follow from such operations.

Additionally, unlike in engineering, in the realm of business, people can be expected to take responsibility for their own actions; they are capable of

---

85. The term "manager" frequently evokes images of middle-level managers, hapless and browbeat, sandwiched between lower-level employees who detest them and upper-level executives on whose behalf they carry out directives. Here, however, the term "manager" is meant to refer to an employee entrusted with carrying out and taking responsibility for the actions of corporations.

rational decisions. Therefore, the responsibility of corporations is not to protect against harms but only not to cause unnecessary harms. According to the role responsibilities of corporations, their duties are limited, in terms of this injunction against producing harms. An exception to this limitation is the workplace itself.

In these environments, corporations have positive duties to protect workers, providing safe and healthy working environments. These duties follow from the fact that employees do not themselves create these environments. Rather, employees are placed in these circumstances. The responsibility to protect employees follows from the decision-making powers of corporations.

6.5.2   Corporations and Fairness: Corporations Should Endeavor to Ensure That All Stakeholders of Their Organizations are Treated Fairly and Justly

Again, the foundation for this claim is a more general ethical principle: justice is a precondition for people to live together in harmony. This is an especially important principle in business, given the previous assumption regarding the dominance of capitalism and the essential role that competition plays in capitalism: competition only produces results beneficial to society when carried out on a fair basis. Bribery, to gain advantage, or bid rigging, to undercut fairness, for example, result in advantages for those involved that are not based on competition, undermining the very foundations of capitalist enterprises. In other words, as established by society, capitalism can only benefit society if it is based on ethical norms.[86]

6.5.3   Corporations and Laws: Corporations Should Endeavor to Ensure That All Relevant Laws and Regulations are Followed Within Their Organizations

In any particular society, laws and regulations establish a framework for business conduct. As was mentioned previously, depending on the society, laws can vary greatly. Businesses have a duty to obey laws, as do all citizens; otherwise, society would cease to function. They are also responsible for the actions of their employees. Since laws are political instruments, as was discussed before, they can run against ethical responsibilities and, for this reason, cannot act as the final arbiters of appropriate ethical actions. For example, corporations are not obliged to follow laws that require discriminatory behaviors. Thus, corpora-

---

86. Although contemporary discourse seems to cast the foundations of capitalism in no-holds-barred terms, even some of the strongest and most well informed proponents of capitalism, such as Adam Smith, Friedrich Hayek, and Milton Friedman, have recognized the necessities of regulating market behaviors. For a discussion along these lines, see, for example, Smith's Words (2013).

tions are required to analyze whether conducting business in societies with such laws is appropriate, balanced against the responsibilities to their investors to seek profits.

6.5.4   Corporations and Discrimination: Corporations Should Endeavor to Protect Members of Their Organizations Against Internal Discrimination and Harassment

This principle is grounded on the more general ethical claim to respect persons, which includes being treated fairly and with respect. Managers have a responsibility to investors to provide the greatest possible financial returns, provided this occurs on an ethical basis. Discrimination based on factors unrelated to work performance results in the waste of talent, thereby decreasing productivity. In the same manner, those exposed to harassment are unable to perform at peak efficiency, thereby decreasing returns on investments.

6.5.5   Corporations and Compensation: Corporations Should Endeavor to Make All Hiring, Compensation, Promotion, and Termination Decisions Based on Merit

At times, managers might have personal likes and prejudices. In acting on these, however, they undermine the interests of investors, for whom managers should act as agents. Managers fail to fulfill their financial responsibilities to investors if they allow personal feelings to interfere with their decisions concerning employees. Additionally, failure to adhere to this principle results in a violation of the more general ethical norm concerning the fair treatment of individuals. It should be noted, however, that this principle would be applied differently depending on the nature of the corporation: if the owners of businesses are the same as their managers, then interests aside from profits can be considered directly. If the owners of businesses are different from their managers—in other words, most corporations—then interests aside from profits are more difficult to determine. Corporate resolutions—such as votes at annual stockholder meetings—allow managers to discover such interests.

6.5.6   Corporations and Contracts: Corporations Should Endeavor to Ensure That All Legitimate Corporate Contracts are Upheld

This principle is based on the more general ethical norm that promises should be kept, which is necessary for people to live and cooperate together in relations of mutual trust. If promises are not kept, then people cannot reach lasting agreements. In business environments, promises generally take the form of contracts, in relation to employees, suppliers, and customers. Although contracts are understood in a more flexible fashion in some cultures and societies, on an international basis, most would agree that contracts should form binding agreements, circumventing the needs of contracting parties to rely on cultural interpretations or tacit understandings:

- Give three examples of how corporations endeavor to keep the public and their employees from harm.

6.6   ETHICAL PRINCIPLES FOR EMPLOYEES

As was mentioned above, since employees occupy positions of subordination in relation to corporations—where corporations are understood as persons—a separate set of ethical principles is necessary for employees. It should be kept in mind that, despite functioning as decision-makers on behalf of corporations, managers are also employees. Even presidents of corporations are ultimately the employees of those who invest in corporations.

6.6.1   Employees and Directives: Corporate Employees Should Endeavor to Obey All Legitimate, Job-Related Directives

Most generally assume that employees enter the employ of corporations based on freely contracted relations. Employees freely serve corporations in exchange for wages or salaries. A requirement of obedience thus follows from the nature of this exchange. Otherwise, hiring individuals would not make sense for corporations. This does not mean, however, that any and all orders should be followed and that any and all job-related demands are legitimate. Directives should be in accordance with the nature of the contracted work. That is the nature of what has been promised. In recent years, increasing recognition exists regarding the need to formalize such arrangements in employment contracts and job descriptions. Requiring employees to perform duties unspecified in their job descriptions is unethical. Furthermore, only legal and ethical demands comprise legitimate aspects of work contracts and job descriptions. For example, employers cannot legitimately demand that their employees commit murder.

6.6.2   Employees and Performance: Corporate Employees Should Endeavor to Perform Their Contracted Duties on at Least an Industry-Standard Level

As mentioned above, in entering employment contracts, employees make promises to employers to perform certain jobs. This raises a question concerning the level at which duties associated with jobs need to be performed to be considered fulfilling contractual obligations. Employers would, of course, prefer that employees devote all of their time and energies to their jobs. However, this is an unreasonable demand. Employees have interests aside from those of work, and doing one's best is not an ethical duty. Employees might be more highly appreciated for doing outstanding work, but doing so is not an ethical requirement. Or, employees might seek additional compensation or career advancements and, thus, do more than is required ethically. That would be their decision. An employment contract only establishes minimally required expectations. Obviously, the phrase "industry-standard level" requires interpretation, given the specific circumstances and cultural conditions in which it is used.

6.6.3 Employers and Confidentiality: Corporate Employees Should Uphold the Principle of Confidentiality in Relation to Knowledge Gained in Present and Past Employment

Private ownership lies at the very basis of capitalism. A right to property thus follows from the previous assumption of capitalism as the particular economic system in terms of which to orient a discussion of business and engineering. In this system, knowledge can be a form of property, generally viewed in the corporate world as a form of "proprietary information." This knowledge can take the form of company secrets or, more openly, the form of patents. The former requires active steps to preserve the secret information, such as restricting access to the information to certain members of the organization. Violating a duty of confidentiality with regard to such information is thus, in effect, like taking property that rightly belongs to someone else, stealing. This applies to both past and present employers, since an employee changing employment does not mean the previous employer no longer owns the information gained by the employee. This fact raises another question regarding the extent to which knowledge gained by an engineer can be used later in life.

Obviously, not everything an engineer learns in the course of employment can be considered confidential. Otherwise, no significant external career advancement would be possible by engineers through changing employers. Requiring that engineers ignore all past on-job learning is obviously unrealistic, although previous employers would like to keep as much information as possible confidential. The use of such knowledge thus requires reasonable judgment on the parts of engineers. Another issue—which will be discussed at length in Chapter 9—is the extent to which engineers might be required to breach confidentiality based on ethical requirements related to public safety.

6.6.4 Employees and Harm: Corporate Employees Should Endeavor to Avoid Actions That Harm the Corporation in Acting on Behalf of the Organization

In working for corporations, employees should benefit those organizations. Otherwise, again, there would be no reason for corporations to employ them.

Additionally, in harming the corporations for which they work, employees, in a sense, also hurt themselves: they violate a more general ethical principle against injuring oneself, acting against their own interests. Finally, employees have a duty of loyalty to their employers.

Again, however, it should be kept in mind that individuals occupy various roles, and the working lives of individuals are not the whole of their lives: individuals also have private lives, and what they do in their private lives has the potential to harm their employers. It is by no means clear, however, that employers are justified in forbidding private actions that run counter to the interests of corporations. For example, should employees of an automobile manufacturer be permitted to drive a car built by a competing manufacturer? Corporations tend to consider all such situations illegitimate conflicts of interests. The extent to which this would be true depends, in part, on other role responsibilities engineers have, which will be further considered later.

6.6.5   Employees and Honesty: Corporate Employees Should Endeavor to be Honest in Their Business Relationships With Others

Workable human relationships require honesty. Without honesty, people are unable to trust each other and, thus, are unable to enter into fully meaningful human relationships. Although largely impersonal, corporate businesses nevertheless involve human relationships and, therefore, require honesty. Furthermore, dishonesty in business relationships undermines productivity—through the inability of dealing effectively with customers and suppliers—and results in the failure of employees to fulfill their duties to corporations.

6.6.6   Employees and Ethics: Corporate Employees Should Endeavor to Ensure That, When in Positions of Authority, They Enforce All Organizational and Employee Ethical Principles

All employees have the responsibility to act ethically, both on practical and ethical grounds. This is an individual responsibility. Those in positions of corporate authority have the additional duty to ensure that the ethical responsibilities of corporations are promulgated and enforced within corporate environments. As has been explained, these responsibilities are directly related to the fiduciary responsibilities of corporate managers, and these responsibilities comprise increasingly significant parts of the roles of modern managers. In many countries, managers can be and are held responsible for the actions of employees. In international environments, future managers should be aware of these responsibilities:

- When should employees uphold the principle of confidentiality with regard to knowledge gained from an employer? Explain your reasoning and give at least one example.

- In the Toshiba Machine Tools case, engineers were closely involved in the sale of the technology. They both demonstrated the technology to the Soviets in secret, after-hour sessions in Japan and installed the milling machines in the Baltic shipyards (Koepp, 1987, p. 53; Triplett, 1988, p. 13). What recommendations do you have for the engineers? Why?
- The Toshiba case relates to strategically important technology, which could be used for military purposes. Does this make any difference in your assessment of the case? If so, how?

6.7 ENGINEERING IN THE CONTEXT OF BUSINESS

Businesses are hierarchical organizations. This is an important feature of their structures. The degree of authority—and consequent responsibility—increases as one moves up the organizational ladder. The authority held by corporate managers is, thus, based on their positions within corporate organizations and is called "institutional authority." This authority is not directly related to competence. Rather, some have argued that employees are promoted in organizations until they reach levels at which they are incompetent.[87]

As corporate employees, engineers hold different positions within organizational structures, including positions with managerial responsibilities. As was discussed in Chapter 3, however, all engineers—regardless of their institutional roles—have "expert authority" based on their specialized knowledge and experience. In effect, this means that engineers have two sets of ethical responsibilities: one based on their roles in institutions and one based on their roles as experts. At times, these roles can conflict. Deciding which role should take precedence then becomes an ethical issue. Resolving this question depends on the circumstances of particular situations. Within the literature on engineering ethics, however, the generally accepted position has been that the role of engineer—and corresponding responsibilities based on expert authority—should take precedence, based on the responsibilities of engineers to protect the safety of the public. The most famous example on which this claim is generally based is that of Bob Lund in relation to the Space Shuttle Challenger disaster.

Bob Lund was the engineering vice president of Morton-Thiokol, the contractor responsible for the rocket boosters on the Space Shuttle Challenger. O-ring deterioration within the booster system ultimately led to the disaster. The night before NASA launched the ill-fated Challenger mission, Lund was told by his superior to "take off your engineering hat and put on your management hat," leading Lund to reverse his earlier recommendation not to launch the Challenger (Martin & Schinzinger, 2010, pp. 98–99). From the point of view of Morton-Thiokol as a corporation, institutional or managerial authority ultimately took precedence over expert authority or advice. Morton-Thiokol

---

87. This is known as the "Peter principle," named after Laurence Peter, who first proposed it.

assumed that final responsibility for decision-making rested with the organization. Despite the horrific consequences of the Challenger disaster, which view should prevail is not always clear.[88]

Engineers should be prepared to have their decisions challenged, often on the grounds that managers have broader perspectives on particular situations. It might be argued that these perspectives take into account factors from outside the realm of engineering, of which engineers might not even be aware. This is especially a danger for young and relatively inexperienced engineers: they tend to view themselves as merely employees bound to the instructions of their institutional superiors, where failing to follows these instructions puts their livelihood at risk.

Additionally, in many cultures and societies, people are raised to defer to the wishes of their elders, thereby tending to obey perceived authority. When placed in rigid organizational structures, they might perceive their engineering roles in relatively narrow terms, where their responsibilities extend to completing technical assignments within pregiven constraints but no further. As has been mentioned, however, blind adherence to these constraints can result in violations of engineering ethical principles.

For these reasons, engineers should learn how to anticipate and appropriately evaluate costs and benefits associated with obedience to hierarchical authority versus exercise of professional autonomy, which will likely be required of them in their working lives as practicing engineers. This can be difficult and, for that reason, is further discussed in Chapters 8 and 9. If one recognizes the potential for conflicting duties, however—learning how to anticipate, navigate, and mitigate any potential conflicts before they arise—then the associated difficulties decrease:

- In the case of the Space Shuttle Challenger, Bob Lund was told to "take off your engineering hat and put on your management hat," leading Lund to reverse his earlier recommendation not to launch the Challenger. How should engineers resolve conflicts between business and engineering responsibilities? Which role do you think should take precedence? Why?

6.8 BUSINESS AND ENGINEERING ETHICS: POINTS OF CONVERGENCE RATHER THAN CONFLICT

Since engineers typically work in business environments, it makes sense to consider engineering ethics in these contexts. If ethical principles associated with these two domains generally conflicted with each other, however, then attempting to specify the relations between the two would be a fruitless endeavor.

---

88. For a fuller treatment of the case of the Space Shuttle Challenger explosion, see Romzek and Dubnick (1987).

Beginning with the reality that most of the time engineers do success-fully function in such environments leads to the conclusion that the principles associated with these two domains not only cohere but also strengthen each other. This conclusion seems reasonable. Were this not the case in a more general sense, then responsibilities associated with the various roles people occupy would constantly conflict, and human beings would be largely dysfunctional creatures. Instead, human beings seem generally capable of integrating the different roles they occupy and prioritizing the responsibilities associated with these roles. Establishing a connection of business with engineering ethics is, thus, only one example of processes in which all people are constantly engaged. Any role ethics has to engage in a similar process, integrating and prioritizing responsibilities associated with various roles. Thus, overlap exists between engineering and business ethics.

First, since neither set of principles are applied independently of specific sets of circumstances and particular contexts—including those of culture and society—neither set of principles should be considered independently from the contexts in which they are applied. More generally, this implies that ethical principles are neither applied nor considered apart from real-life constraints. Since both business and engineering ethics principles are understood as flexible, based on the processes of their applications, this provides a point of convergence between different forms of applied ethics. As was mentioned before, this justifies the centrality of studying specific cases.

Next, although business and engineering ethics share common concerns—such as honesty and legality—to a significant extent, they deal with different domains and associated issues. For example, business ethics is not particularly concerned with the implementation of new technologies, and responsibilities within engineering ethics are toward the general public rather than shareholders or stakeholders specifically. If this were not the case, then there would be no need for two sets of ethical principles. For that reason, these two forms of ethics need not conflict; they have their own respective domains of concern.

Finally, different forms of applied ethics relate to each other. The principles of engineering ethics are derived from the nature of engineering itself. Engineering is a human activity and, as such, it is based on human interactions. It shares this fundamental feature with business. That means a set of common norms governing human interactions, in general, stands at the core of both engineering and business. Their application across different realms shows that similar ethical principles function in both. In this regard, even a cursory examination of the ethical principles examined thus far reveals significant overlap between the fields of business and engineering ethics.

The avoidance of harm, for example, is emphasized in both sets of principles. This is unsurprising, since avoiding harm to others is a general ethical concern likely to arise in any field of applied ethics. This is a reminder that, fundamentally, ethics concerns the interests of others rather than simply self-interests. For this reason, competence is another commonality within activity-based ethics.

Self-interests might dictate a desire to benefit regardless of the quality of the consequences of activities. Once the interests of others are considered, however, then competent performance is required in a wide range of human activities. As a teacher, for instance, one might wish to be paid regardless of the quality of one's teaching, but if the interests of students are considered, then how much students learn becomes relevant. Honesty is a final example of a commonality between principles of ethics in different domains.

Honesty—the provision of true information—stands at the foundation of all fair human interactions. For this reason, honesty should figure prominently in most forms of applied ethics. If teachers lied to students about the requirements to pass courses, for instance, then students would not be in positions to make informed decisions about their actions. These are only a few examples of points of commonality between the two sets of ethics discussed above. More detailed examinations by readers would reveal many more:

- Do you agree with the claim that the principles associated with the ethical domains of engineering and business "not only cohere but also strengthen each other"? Why or why not?

6.9   POTENTIALS FOR CONFLICTING DUTIES: POINTS OF CONFLICT RATHER THAN CONVERGENCE

Although overlap exists between business and engineering ethics, the potential for conflicting duties should also be recognized. Again, if this were not the case, then there would be little need to examine these two fields of ethics separately; lists of duties applicable to all employees could simply be applied to engineers, and analysis would be complete. In considering potential conflicts, the most important point to keep in mind is that the bases for making decisions in these two domains are different: decisions in business are based primarily on profits, whereas decisions in engineering are based primarily on the implementation of technologies. Based on this distinction, several sources of potential conflict can arise.

As technical experts confronted by a relatively ignorant public, engineers have a positive duty to protect people from physical harm. In business, the assumption is that people are capable of making their own decisions. Therefore, ethical obligations within business are limited to not creating harm. The stronger duty within engineering can, therefore, potentially conflict with the corporate duty of securing profits, since protecting the public generally requires financial resources. Confidentiality with regard to proprietary information is another area in which potential conflicts can arise.

As was discussed previously, on the one hand, corporations have legitimate interests in preserving information for their own uses, thereby increasing the potentials for profits. On the other hand, in introducing technology into the world—based on the analogy of social experimentation—engineers are committed to openness in obtaining informed consent from the public as far as is possible. A final example of potential differences in emphases between business and engineering ethics concerns the groups to whom ethical responsibilities should ultimately be directed.

On the stakeholders view, although corporations should take into account the interests of all affected parties in making decisions, the interests of all parties should not receive equal weight. On the one hand, from a corporate point of view, the satisfaction of investors and customers is primary. On the other hand, from an engineering point of view, the safety of the public as a whole is primary. In working for corporations, resolving the issue of which parties have priority can become a moral dilemma for engineers.

The purpose of this section is not to resolve potential conflicts. That requires more detailed analyses. The intention is merely to note that engineers should recognize the potential for such conflicts. In addition to direct conflicts, the ranges of duties between business and engineering ethics differ significantly, as does the process of establishing obligational hierarchies between the two.

Finally, although this section has discussed the potential for conflicts between principles from different branches of applied ethics, the potential for conflicts between principles from one domain of applied ethics—within engineering ethics alone, for instance—also exists. Resolving such conflicts cannot be decided in the abstract, based on a more general ethical principle, but should be decided on a case-by-case basis. Once again, this highlights the central importance of the case-study procedure to applied ethics in general, and the importance of students not only learning about but also practicing the application of the case-study procedure.

EXERCISE TWO—THE CASE OF TOSHIBA MACHINE TOOLS (PART TWO)

Return to the second step of the case study procedure in the exercise you completed at the beginning of this chapter, where you identified the most important ethical issue in the case. Using the ethical principles for organizations and employees discussed in Sections 6.6 and 6.7, respectively, complete step 6 of the case-study procedure again. In addition to the basic ethical principles for global engineering you identified before, list any additional principles for ethical organizations and employees that would be relevant to resolving the issue you identified. If conflicts exist between these principles, then establish a hierarchy between these duties, providing a brief justification for the hierarchy you chose. Do you think this ranking in the principles should always apply or only in this case? Again, provide a brief justification for your response.

CASE STUDY TWO—VOLKSWAGEN, INTERNATIONAL BUSINESS, AND THE ENVIRONMENT

Introduction: Consequences and Ethics in Business

On Sep. 18, 2015, the US Environmental Protection Agency (EPA) concluded that the German automobile manufacturer Volkswagen (hereafter VW) violated the Clean Air Act and could be fined as much as $37,500 per vehicle as a result—in the United States alone, "up to $18 billion" (Hotten, 2015; Makortoff, 2015)—although millions of cars were affected worldwide. In Jan. 2016, the US Department of Justice sought close to $48 billion in penalties (Hotten, 2015). Since then, VW has agreed to a number of settlements totaling billions of dollars (Shepardson, 2016).

The charges related to software known as "defeat devices" installed to trick tests measuring nitrogen oxide emissions from VW diesel automobiles. Hazardous emissions from the VW cars with the software were 15–35 times higher than those measured by the fixed emissions tests (Taebi, 2015). In the aftermath, VW's stock dropped by 50%—from Mar. of 2015 to 2016—(Hotten, 2015) and Martin Winterkorn, the CEO of VW, resigned; Matthias Mueller, the President and CEO of Porsche replaced him. As a result of the software, the *Guardian* newspaper estimated that VW cars could be "responsible for nearly 1 million tonnes of air pollution every year, roughly the same as the United Kingdom's combined emissions for all power stations, vehicles, industry and agriculture" (Mathiesen & Neslen, 2015). On Oct. 6, the VW Group of America returned three Cars.com awards it had previously received for clean diesel vehicles (Mays, 2015). This case brings together issues at the center of global engineering, international business, ethics, and the environment.

Despite its origins in National Socialist Germany, VW had enjoyed a reputation for producing reliable vehicles, supporting the national engineering reputation of Germany and Germans as honest and diligent workers. A corporate ethos and reputation such as this undoubtedly contributes to the success of companies in global business, increasing revenues and profits worldwide. From this perspective, ethics makes sound business sense: companies can and do succeed in business precisely by being ethical. As the effects of human behavior on the environment become better understood, corporations that fail to adjust their business accordingly not only risk incurring legal and economic sanctions but also lose out on business that takes seriously these concerns.

Background of the Case: The Software, Vehicles, and Discoveries

Since 2009, VW cars in the United States outfitted with defeat devices "produced doctored results" when tested for nitrogen oxide emissions ("Volkswagen Chief," 2015). Analyzing "the position of the steering wheel, vehicle speed, the duration of the engine's operation and barometric pressure" (Makortoff, 2015), the software was capable of determining whether the vehicle was in testing conditions, running a "clean program" that would reduce power and resulting emissions (Taebi, 2015). When operating normally however—driving on the road—the cars produced "emissions of nitrogen oxide up to nine times EPA's standard" ("United States," 2016).

The software had been installed in millions of vehicles: "8.5 million cars in Europe, including 2.4 million in Germany and 1.2 million in the UK, and 500,000 in the US" (Hotten, 2015). These were diesel-engine models, including Volkswagen Jettas, Beetles, Golfs, Passats, and Touaregs, as well as Audi A3's, A6's, A7's, A8's, and Q5's, and Porsche Cayennes (Brooks, 2015; Shinkman, 2015). Questions regarding the emissions of VW automobiles began as early as 2012.

From late 2012 to May 2013, supported by a grant from the International Council on Clean Transportation, engineers from West Virginia University determined "Volkswagen was cheating on US vehicle emission tests" (Kim & Pickering, 2015). They did so using on-road emissions tests, rather than those generally used, in which the car is tested while stationary. From May 2014, the California Air Resources Board (henceforth CARB) followed up, also testing VW vehicles. VW was notified of the testing and problems around this time (Lam, 2015).

VW's Response: Disingenuous and International

Once the scandal received public attention, the CEO, Winterkorn, said in a company statement that he was "shocked by the events of the past few days"— although there were indications Winterkorn knew of and ignored issues related to VW emissions going back some time (Makortoff, Boyle, & Tutt, 2015). After being alerted of the test results, VW claimed "higher nitrogen oxide emissions" were caused by "technical glitches" (Boston, Spector, & Harder, 2015), telling CARB the problem was caused by "unexpected in-use conditions" (Brooks, 2015)—in general, dismissing the findings as resulting from "technical issues and 'unexpected' test conditions" (Mays, 2015). Moreover, VW began installing defeat devices in their 2009 models, once Winterkorn was already CEO for 2 years ("Volkswagen Chief," 2015). As a German company, reactions to the VW scandal were different in Germany and the United States.

German VW executives faced less criticism. In response to the revelations, German regulators "made only the briefest of statements on the scandal, preferring to focus on how to fix the problem of the vehicles" (Copley, 2015). The German Association of the Automotive Industry (Verband der Automobilindustrie)—representing Germany's and perhaps the world's foremost car companies—remained relatively silent on the matter. Additionally, the Kraftfahrt-Bundesamt (KBA—Federal Motor Transport Authority of Germany) appeared to publicly support VW, insisting its actions called for no serious penalties: "Spokesman Stephan Immen said the KBA had no precedent of imposing penalties on car manufacturers for wrongdoing" (Copley, 2015). As with the cases of Ford and Firestone/Bridgestone and Toshiba Machine Tools, reactions in the United States were different. Michael Horn—responsible for VW US operations—bluntly stated that VW "totally screwed up," admitting the fault of VW and promising to fully cooperate with the upcoming investigation and future regulations (Makortoff, 2015):

- Which companies and/or brands with which are you familiar have reputations as being particularly ethical or unethical? Why do they have these reputations? Which characteristics, actions, or events contributed to these reputations?
- Why do you think reactions to the VW scandal were different in Germany and the United States? Describe any potential conflicts of interest you think might have played a role in this difference.

EXERCISE THREE—VOLKSWAGEN, INTERNATIONAL BUSINESS, AND THE ENVIRONMENT

Complete the case-study procedure on Volkswagen, International Business, and the Environment using all the ethical principles introduced and discussed up to this point.

6.10   SUMMARY

As becomes clear by examining the Toshiba Machine Tools case, contemporary engineering practices rarely take place in isolation. For this reason, it is important to examine the environments, corresponding roles, and consequent duties related to contemporary engineering practices. Chief among these are the global business environment and the roles of engineers as employees. As with engineering ethics in global contexts, to better understand the relations between engineering and business, it is necessary to make general descriptive assumptions regarding the natures of contemporary business practices and environments. As with engineering, a number of principles for business ethics follow from these assumptions. These principles can be divided and explained in terms of duties applicable to organizations and those applicable to employees. As would be expected—given their mutual reference to human relations—significant overlap exists between principles of ethical engineering and business. At the same time, however, one should be cognizant of the potential for conflicting duties between these sets of principles. The point here is not to give priority to one set of duties or a particular principle in the abstract, but to be cognizant of specific points at which duties and principles have the potential to conflict. The process of establishing a hierarchy between duties and principles should occur through their application to particular cases, highlighting the importance of case studies. As the case of the Volkswagen emissions scandal exemplifies—albeit in a negative fashion—ethical engineering and the interests of business need not conflict:

ethical practices can contribute to positive corporate and national reputations, facilitating the business objective of increasing profits.

REVIEW QUESTIONS

1. List the parties involved in the contract to deliver the eight computer-controlled milling machines and their associated equipment.
2. The investigative report published by the *Detroit News* had repercussions in both the United States and Japan. How were these different in the two countries?
3. How did the MITI help Toshiba Machine avoid COCOM regulations? What effect did these actions have on the conflict between the United States and Japan?
4. List and explain the three assumptions made regarding the nature of global business.
5. List and explain at least two differences between the ethical obligations of managers and the ethical obligations of their subordinates.
6. Describe the foundations on which business ethical principles are established and their difference from engineering ethics.
7. What are the most important points to keep in mind when considering potential conflicts between engineering and business ethics?
8. Contrary to some claims, why should ethics apply to the realm of business? Provide at least two justifications and relevant examples.
9. Explain the difference between the shareholders and stakeholders perspectives, and why these definitions are significant to business ethics.
10. List the immediate and potentially long-term business consequences to VW of the emissions scandal.
11. Describe the nature of the defeat devices VW installed in their diesel vehicles—how they managed to trick the vehicle emissions tests.

REFERENCES

Armstrong, L., et al. (1987). The toshiba scandal has exporters running for cover. *Business Week,* July 20.
Boston, W., Spector, M., & Harder, A. (2015). Volkswagen scandal pressures ceo. *Wall Street Journal,* 23 September. http://www.wsj.com/articles/volkswagen-scandal-pressures-ceo-1442967027.
Brooks, P. (2015). Notice of violation. *EPA,* 18 September. https://www.epa.gov/sites/production/files/2015-10/documents/vw-nov-caa-09-18-15.pdf.
Carr, A. (1968). *Business as a game.* New York, NY: New American Library.
Chandler, C. (1987). Bright lights, big MITI-behind the Toshiba scandal. *The New Republic,* 31 August.
Chandler, A. (1990). *Strategy and structure: Chapters in the history of the industrial enterprise.* Boston, MA: MIT Press.
Cook, W. (1987). An illegal deal's noisy repercussions. U.S. News & World Report. 15 June.
Copeland, J. (1987). The battle over Toshiba-corporate Hara-Kiri. *Time,* 13 July.

Copley, C. (2015). Germany rallies around Volkswagen in diesel emissions scandal. *Reuters*, 15 October. http://uk.reuters.com/article/uk-volkswagen-emissions-germany-regulati-idUK-KCN0S91AS20151015.

Davis, J. (2013). *Essays in the earlier history of American corporations*. Clark, NJ: The Lawbook Exchange Ltd.

Dryden, S. (1987). How Toshiba is beating American sanctions. *Business Week*, 14 September.

Eisenhardt, K. (1989). Agency theory: An assessment and review. *Academy of Management Review*, *14*(1), 57–74.

Goldman, M. (1987). *The case of the not-so-simple machine tools*. Technological Review October.

Hard Pounding. (1987). *The Economist*. 11 July.

Hasnas, J. (1998). The normative theories of buisness ethics: A guide for the perplexed. *Business Ethics Quarterly*, *8*(1), 19–42.

Hotten, R. (2015). Volkswagen: The scandal explained. *BBC*, 4 November. http://www.bbc.com/news/business-34324772.

Kapstein, J. (1987). A leak that could sink the U.S. lead in submarines. *Business Week*, 18 May.

Kim, S., & Pickering, J. (2015). How West Virginia engineer exposed Volkswagen's catastrophic environmental fraud and wiped billions off company's shares. *Dailymail*, 23 September. http://www.dailymail.co.uk/news/article-3245167/West-Virginia-engineer-proves-David-VWs-Goliath.html.

Koepp, S. (1987). Beware of machines in disguise. *Time*, 21 September.

Lam, B. (2015). The academic paper that broke the Volkswagen scandal. *The Atlantic*, 25 September. http://www.theatlantic.com/business/archive/2015/09/volkswagen-scandal-cheating-emission-virginia-epa/407425/.

Making Amends-Top Toshiba Executives Resign. (1987). *Time*. 13 July.

Makortoff, K. (2015). What you need to know about the Volkswagen scandal. *CNBC*, 22 September. http://www.cnbc.com/2015/09/22/what-you-need-to-know-about-the-volkswagen-scandal.html.

Makortoff, K., Boyle, C., & Tutt, P. (2015). Martin Winterkorn resigns as Volkswagen CEO. *CNBC*, 23 September. http://www.cnbc.com/2015/09/23/martin-winterkorn-resigns-as-volkswagen-ceo.html.

Martin, M., & Schinzinger, R. 2010. *Introduction to engineering ethics* (2nd ed.). New York, NY: McGraw Hill.

Mathiesen, K., & Neslen, A. (2015). VW Scandal Caused Nearly 1 m Tonnes of Extra Pollution, Analysis Shows. *The Guardian*, 23 September. http://www.theguardian.com/business/2015/sep/22/vw-scandal-caused-nearly-1m-tonnes-of-extra-pollution-analysis-shows.

Mays, K. (2015). VW Diesel Crisis: Timeline of Events. *Cars.com*. 23 September. https://www.cars.com/articles/vw-diesel-crisis-timeline-of-events-1420681251993/.

Norway is Selling Kongsberg's Assets to 13 Other Firms. (1987). Aviation Week & Space Technology. 23 November.

Packard, G. (1987). The coming U.S.-Japan crisis. *Foreign Affairs*, *66*(2) Winter.

Romzek, B., & Dubnick, M. (1987). Accountability in the public sector: Lessons from the challenger tragedy. *Public Administration Review*, 227–238.

Shapiro, S. (2005). Agency theory. *Annual Review of Sociology*, 263–284.

Shepardson, D. (2016). U.S. judge approves $14.7 billion deal in VW diesel scandal. *Reuters News*. 25 October. https://www.epa.gov/sites/production/files/2015-11/documents/vw-nov-2015-11-02.pdf.

Shinkman, S. (2015). Notice of violations. *EPA*, 2 November. https://www.epa.gov/sites/production/files/2015-11/documents/vw-nov-2015-11-02.pdf.

Smith, J. (2003). The shareholders vs. stakeholders debate. *MIT Sloan Management Review*, *44*(4), 85–91.

Smith's Words (2013). *Economist*, 1 November. http://www.economist.com/blogs/freeex-change/2013/11/economic-history?fsrc=scn%2Ffb%2Fte%2Fbl%2Fed%2Feconomichistorys mithsword.

Taebi, B. (2015). Volkswagen scandal reveals design flaws: The need for value conscious design. *The Huffington Post*, 9 November. http://www.huffingtonpost.com/behnam-taebi/volkswagen-scandal-reveal_b_8280708.html.

Toshiba Ban Spurs Review of Technology Export Controls. (1987). *Aviation Week & Space Technology*. 6 July.

Toshiba vs. the U.S.: A Japanese View (1987). A. Shimbum (trans.). *World Business Review*. August.

Triplett, W. (1988). Crimes against the alliance—The Toshiba-Kongsberg export violations. *Policy Review*, *44*, Spring.

United States Files Complaint Against Volkswagen, Audi and Porsche for Alleged Clean Air Act Violations. (2016). *EPA*. 4 January. https://www.epa.gov/newsreleases/united-states-files-complaint-against-volkswagen-audi-and-porsche-alleged-clean-air-act.

Volkswagen Chief Executive Martin Winterkorn Resigns. (2015). *BBC*. 23 September. http://www.bbc.com/news/business-34340997.

FURTHER READING

Fuel on the Fire. (2015). *The Economist*. 7 November. http://www.economist.com/news/business/21677623-another-blow-german-carmaker-fuel-fire.

Chapter 7

Cross-Cultural Issues: Their Importance to Global Engineering Ethics

Chapter Objectives

Having read this chapter, completed the included exercises, and answered the associated questions, readers should be able to

- with reference to the case of bid rigging in Japan, identify and explain the significance of cultural values and their relation to ethical behaviors;
- describe why it is important for engineers to be aware of cultural values when working in international and cross-cultural environments;
- explain "normative ethical relativism" and why is it an unsatisfactory position for an approach to engineering ethics in global contexts;
- show an understanding of the nature and importance of the distinction between moral and nonmoral cultural values, as well as why this distinction should be important to engineers.

CASE STUDY—BID-RIGGING IN JAPAN: AN ETHICAL OR CULTURAL ISSUE?

"Bid rigging" is when parties competing for a contract collude in secret to decide the winner—a form of "price fixing" aimed at insuring greater profits. Although bid rigging occurs throughout the world, in Japan it has received a great deal of attention. Focusing on this issue brings together engineering and business—examined in previous chapters—and concerns regarding the nature of cultural values, touched on before and dealt with at length in this chapter.

Dangō, the Government, and Law

In Japan, bid rigging is called *dangō*, a combination of the words for "discuss" and "meet." Unlike its English equivalent, the term does not necessarily have a negative connotation: cooperative activities by firms in the same industries have often been encouraged in Japan. Various methods exist for allocating the profits gained from bid rigging, which include rotating the winners and dividing

Global Engineering Ethics. http://dx.doi.org/10.1016/B978-0-12-811218-2.00007-2
125

the profits. For bid rigging to be successful, John McMillan has argued that five conditions are necessary: "a means for determining the composition of the conspiratorial ring (1), barriers to prevent opportunistic outsiders from undercutting its action (2), informal devices to ensure compliance (3), a mechanism for dividing the spoils (4), and some means of evading or co-opting government watchdogs (5)" (McMillan, 1991). In public works projects in Japan, he argues, all of these conditions are met.

Kansei dingo is a specific form of bid rigging where government officials or agencies are involved, by either revealing confidential information to bidders or guiding the allocation of funds. In most industrialized countries, bid rigging is illegal. It is illegal in Japan—in terms of antimonopoly legislation—and criminally prosecutable. A law was passed in 2003 to punish government officials involved in bid rigging called the "Act Concerning Elimination and Prevention of Involvement in Bid Rigging."

Background in Japan

Before World War II, the Japanese government encouraged the development of cartels to strengthen the country's heavy industries. Cartels are based on collusive activities forming the background of contemporary bid-rigging practices. In 1947, Japan passed antimonopoly legislation, although it has been subject to a variety of exceptions over the years (Iyori & Uesugi, 1994). Since that time, bid rigging has become a common practice in the construction industry and in other situations involving government contracts.

In approximately 90% of the Japanese public works contracts open for bidding, the following system is used: the awarding agency ranks bidders based on past performance, company size, technological capabilities, or other criteria, and then develops a list of eligible bidding participants. This is known as "discretionary bidding," since participation is based on the discretion of the government agency. This is in contrast to competitive bidding, where all potential participants are allowed to bid, provided they meet a minimum set of criteria. Discretionary bidding has been used to favor local companies and small- and medium-sized firms. If government officials are involved in bid rigging, then they can disqualify noncooperating firms from future competitions—firms that do not collude in fixing prices.

Amakudari—meaning "descent from heaven"—is another feature of the Japanese bureaucratic system relevant to bid rigging. In Japan, government officials are required to retire at a relatively early age, after which time many take up positions in the industries they previously regulated, thereby securing additional income and generous retirement allowances. Some *amakudari* positions are at semipublic agencies and foundations that participate in the bidding process as either the awarders or awardees of bids.

Many have suspected the involvement of government officials in bid rigging, including police officers, city mayors, prefectural governors, and central ministers. In some cases, it is believed that they receive direct kickbacks, in other cases, promises of future jobs or contributions to their political organizations. In Japan, the construction industry is the biggest financial contributor to political parties. It also spends lots of money "wining and dining" government officials.

Cultural and Ethical Issues in Bid-Rigging

Bid rigging is often only considered in the context of business ethics. However, as it has the potential to affect the quality of work performed and, therefore, public safety, it has implications and consequences for engineering ethics.

One issue concerns the relations between bid rigging and Japanese values: first, as it requires cooperation between firms in an industry, bid rigging evidences the values of harmony and compromise. Second, the practice places an emphasis on a group of firms rather than any one individual firm. Third, bid rigging ensures the smooth operation of the bidding process and enables the survival of a group of firms as a whole. Finally, *amakudari* establishes a web of relations between the companies and the government. Thus, it can be asked whether a natural association exists between bid rigging and more general Japanese cultural values.[89]

A second issue concerns the relations between bid rigging and bribery: the latter has been of major interest within the field of traditional business ethics. Does bid rigging necessarily entail compromising the integrity of government officials? The Japanese government typically establishes price ceilings for public works auctions that could be well above typical cost-plus-profit calculations. Bid riggers have typically been aware of these price ceilings and have submitted winning bids within a few percentage points. Those bidding have obviously obtained this information from sources close to the projects. It can be asked whether the implicit promise of a later job or political contributions constitute a form of bribery.

A third issue concerns the relations between bid rigging and competition: some have argued that bid rigging leads to lower overall quality and compromises safety due to a desire to produce the lowest possible bid. Others have argued that the lack of competition that follows from bid rigging has this same effect, since corporations need not be tied to the quality of their work to receive future contracts.

A fourth issue concerns the relation between bid rigging and social costs: one can assume that the costs of public works projects involving bid rigging will be higher than those not involving bid rigging. Even if this process results

---

89. For surveys regarding the nature of Japanese values and society, see Nakane (1970) and Smith (1983). Concerning the way conflicts between values related to security and safety have been important to the nuclear power industry in Japan, see Luegenbiehl (2009).

in greater social harmony and higher quality products, one can still ask whether these benefits outweigh the economic costs. Further, some have claimed that processes of bid rigging and their connection to the government result in projects that are unnecessary in the first place and, thus, unwarranted economic costs to society.

A final issue concerns the relation between bid rigging and cross-cultural contexts: even if a particular society is willing to bear the costs associated with bid rigging for reasons related to overall social wellbeing, its relation to cross-cultural contexts is unclear. For example, in the 1980s, it was discovered that bid rigging was occurring in the process of obtaining contracts for US naval bases in Japan. Additionally, in the construction of the Kansai airport, foreign contractors were excluded from bidding on a discretionary basis. The latter became the basis for extensive trade friction, where the claim was made that this system gave rise to informal trade barriers. Assuming an acceptance of these practices within Japan, what if several Japanese companies colluded to rig bids abroad?

As a result of bid rigging, Japanese government officials and corporate executives have been arrested—although they have not necessarily received heavy punishments—and the public has seemed genuinely upset with the perpetrators. However, the phenomenon has not ceased. Even when industries have been punished, they have continued to engage in bid rigging. Thus, the issue is clearly a complex one. Although the focus here has been on bid rigging in Japan, as indicated above, this issue is of broader concern. Insofar as Japanese companies operate in international environments—and bid rigging and related practices occurs in industries in other countries as well—the practice clearly deserves further consideration.

EXERCISE ONE—BID RIGGING IN JAPAN (PART ONE)

With regard to bid rigging in Japan, complete the case-study procedure on bid rigging in Japan, using the principles of global engineering, organizations, and employees.

7.1 THE IMPORTANCE OF CULTURAL VALUES WITHIN ENGINEERING

In taking a global approach to engineering ethics, using any one cultural perspective has been avoided. However, a difference exists between establishing a theoretical framework for engineering ethics and the real-life practice of engineering. With regard to the latter, it is important to recognize that local cultures can have significant impacts on engineering practices—especially in the context of business—and engineers should respect local customs and traditions. Therefore, this chapter investigates in greater depth the relation between engineering ethical principles and cross-cultural values.

7.2   NORMATIVE ETHICAL RELATIVISM: IT'S ALL THE SAME

Normative ethical relativism is a significant tradition within ethical theory. This position begins with the factual thesis that different individuals and groups subscribe to different ethical positions—a relatively uncontroversial claim—but makes the further assertion that individuals and groups are justified in subscribing to the beliefs they do. This normative dimension of ethical relativism has elicited a variety of objections, which will not be discussed here.[90] The approach to engineering ethics advocated here does not fall under the category of normative ethical relativism.

The reason for this is as follows: within this tradition, the only *justification* needed for adopting a particular ethical position is the *fact* that individuals or groups adopt it. Individuals or groups might adopt particular ethical positions for good, bad, or no reasons at all. In developing a global approach to engineering ethics, the justifications for the ethical positions adopted here are based on the nature of engineering and use of reason. The process of deriving engineering ethical principles is, thus, nonarbitrary.

Issues related to cultural values are important to consider since, in the 21st century, there are areas in which cultural differences play major roles, and these should be recognized and accepted by engineers. In encountering cultural traditions at variance with their own, engineers should be prepared to decide the circumstances in which to respect these, and the circumstances in which to either refuse to go along with—or actively resist—prevailing cultural trends. To establish a framework for making decisions such as these, it is necessary to examine the nature of values.

7.3   THE NATURE OF VALUES AND CROSS-CULTURAL CONTEXTS

"Values" can be understood as referring to deeply held, enduring, and important beliefs that tend to guide the actions of individuals and groups. Given their deeply ingrained nature, people tend to overlook the central role that values play in decision-making processes. People tend to believe that others share these views, and—in many cases—that values are "natural" or innate.[91]

A distinction can be made between individual and social values. Since every person is, to some extent, different, every person has his or her own set of beliefs. However, insofar as groups of individuals share common backgrounds, it is possible to make broad generalizations. These generalized, action-guiding beliefs are called "social values."

Values do not exist on their own or in the abstract. Rather, they exist in relation to other values and actions as parts of wholes. These wholes are called "value

---

90. For more on normative ethical relativism and objections to this position, the interested reader can consult Rachels (2011).
91. For a broad overview regarding the nature of values in general, see, for example, Rokeach (1973).

systems." Within values systems, individual values are prioritized. Over time, this prioritization can change, reflecting processes of change within individuals and societies. In radical situations, values might be added to or eliminated from value systems but, more typically, values become emphasized and deemphasized over time. Understanding the value systems of particular cultures is important, since values have "normative" dimensions—those regarding judgments.

Values establish expectations regarding behaviors, which are reflected in the traditions, customs, and manners of cultures, in turn establishing parameters of appropriate behaviors. Common social values create unity within populations. A method of differentiating cultures is on the basis of their respective value systems. Although making generalizations about cultures can result in negative effects associated with stereotyping and prejudice, it is important to recognize that, in general, cultures have individuating characteristics. Most cultures share many values, but individual values are integrated hierarchically, such that the importance placed on particular values distinguishes one value system from another. Two different value systems might both include "order" and "freedom," for example, but freedom would usually be given preference over order in one of them, while the opposite might be the case in the other. The complex and hierarchical nature of value systems can give rise to problems.

At times, it can be difficult to determine the value framework underlying the words and actions of individuals from different cultures. However, in these situations, understanding this framework is important. As one's own cultural values are deeply ingrained, one can fail to recognize their roles in decision-making processes. Erroneously, people tend to believe everyone shares their values and uses the same value system, or should do so. Consequently, when individuals from different cultures arrive at different decisions, they tend to believe the other has made a mistake or—in a stronger sense—done something unethical. A more likely explanation, however, is that the two individuals are employing different value systems to guide their decisions and actions.

Insofar as the value systems of some societies are closely aligned, situations of cultural misunderstandings are infrequent, although perhaps more problematic, since they are unexpected. People from the Midwest United States might be surprised to discover, for example, that those from the Northeast are more direct or forthright in their assessments of others. In other cases, however, a large gap exists between value systems. Many have claimed this is the case with "Western" and "Asian" societies, which tend to emphasize different values: in general, Western societies emphasize values associated with furthering the rights of individuals, while Asian societies emphasize values associated with furthering the interests of groups. This difference affects the entire range of values within a system. For example, loyalty to an extended family member or employer would be manifested differently in these two types of societies.[92]

---

92. On these differences, see, for example, Sigurosson (2014) and Garcia, Mendez, Ellis, and Gautney (2014). However, others have argued that these differences are not as great and/or important, see, for example, Caney (1999) and Shafer-Landau (2003).

When encountering individuals from different value systems, one's first re-action might be to reject his or her value system as it conflicts with one's own deeply held beliefs. One might judge the values of others—reflected in their actions—as wrong and seek to correct their seeming ignorance. However, such an approach not only reflects "imperialist" assumptions regarding relations between societies—assuming one's own society and its associated values are fundamentally superior to others—but also undermines the possibility of posi-tively interacting with people from different cultures. Societies should be free to develop their own cultural norms, and it is important for outsiders to learn to respect the values of a culture. However, at least some basis for questioning the values of others exists:

- Why do you think people commonly judge and/or reject values systems dif-ferent from their own? Explain an instance in which you have been inclined to do so.

7.4 VALUES AND ETHICS: MORAL AND NONMORAL VALUES

The connection between values and ethics is an intimate one. Values tend to guide actions, and ethics helps to evaluate actions, although only certain types of actions. In a significant sense, ethics is a subdiscipline of "value theory": values are typically divided into moral and nonmoral ones, where ethics is con-cerned with the former. Both are concerned with making judgments, although judgments based on nonmoral values are not generally considered either right or wrong. In moral judgments, by contrast, claims are made regarding the right-ness and wrongness of decisions and actions.

For example, although two people might like and dislike the same painting, respectively—in other words, they have different "tastes" in art—neither is in a position to question the other's right to like the painting, since preferences in art are based on particular sets of nonmoral aesthetic values. Preferences regarding art and music vary not only from culture to culture but also from individual to individual. By contrast, if the person who likes the painting decides to take it without permission, then that would be theft, and the other would be in posi-tion to judge his or her actions as morally wrong. In a sense then, ethics limits justification of the exercise of nonmoral value preferences.[93]

---

93. Regarding an understanding of the relation between value theory and ethics in these terms, see, for example, Singer (2011). For an explanation of this view in the specific context of Kantian ethics, see Nell (1975). Not all would agree that the disconnection between moral and nonmoral values is as clear as it has been described here. 19th- and 20th-century forerunners to/founders of value theory, such as Friedrich Nietzsche and Karl Marx, and pragmatists such as John Dewey, William James, and Alain Locke have argued for the intimate connection between the two: criteria for mak-ing (nonmoral) judgments concerning aesthetic preferences, for example, mutually and reciprocally determine criteria for making (moral) judgments concerning ethics, politics, etc.

Explicitly recognizing this distinction between moral and nonmoral values is important since, within a given cultural tradition, there might be little or no awareness of it. If this is the case, then all value judgments—both moral and nonmoral—are understood as having equal claims on the individual. All value judgments would be justified on the same basis, as following from the cultural traditions of a given society. Little or no distinction is made between actions that have ethical and nonethical import—in other words, a distinction between actions that have the potential to seriously affect the lives of others and actions that express mere cultural preferences. Since the authority of ethical directives is great, by associating cultural preferences with ethics, nonmoral cultural preferences can take on greater normative force within certain social contexts and are sometimes punished as severely.

Removing one's shoes indoors might be a cultural practice. Failure to remove one's shoes would indicate a failure to understand or respect the values of that society, perhaps cleanliness. However, this failure should not be considered an immoral action, although it might be in the context of that society, due to the amalgamation of moral and nonmoral values. Conversely, at times actions with ethical import are justified merely on the basis of reflecting cultural practices and, for this reason, some would claim, excluded from the sphere of judgment by those from outside the given culture. This is, of course, the position of normative ethical relativism, judged to be an inadequate approach to engineering ethics on a global basis.

In short, it is necessary to distinguish between nonmoral and moral cultural values. The first concerns cultural preferences, and it is important to try to understand these when interacting with people from cultures other than one's own. The second concerns questions of right and wrong—with universal characteristics—and are not simply matters of preference. For example, on the one hand, stealing is generally considered wrong in all cultures, since it leads to a breakdown in social structures. However, the specific nature of stealing—what one considers stealing—might vary from society to society, based on differing cultural interpretations. On the other hand, the extent to which characteristics such as generosity or frugality are valued might vary significantly from culture to culture. The nature of this distinction can be further clarified by comparing two value systems, those of Japan and the United States.

To begin, it is important to remember that values occur in systems, and the places of individual values within hierarchies determine their roles in cultures: as most societies share many values, the emphasis given to particular values distinguishes one culture from another. In general, Japanese society tends to emphasize group-oriented values, while US society tends to emphasize individual-oriented values. Thus, Japanese culture emphasizes values such as harmony, loyalty, hierarchy, consensus, duty, and conformity, while US culture emphasizes values such as rights, freedom, equality, independence, and choice. Both cultures emphasize values such as education, material well-being, and pragmatism. Additionally, both cultures emphasize other values perhaps not directly related to the distinction between the group and individual, for example, ritualism and sincerity in Japan, and religiosity and merit in the United States.[94]

Understanding the value structures of different cultures and their differences from one's own requires detailed acquaintance. Gaining this acquaintance is vital to not only interpreting adequately and accurately the actions of others but also distinguishing between the ethical and nonethical dimensions of actions:

- Explain an incident of which you are aware where an engineering or technology firm failed to take into account moral or nonmoral cultural considerations. What were the consequences?

7.5 VALUES AND ENGINEERING ETHICS: TWO POINTS TO KEEP IN MIND

The following two points have been established, which are important for engineers to keep in mind:

First, the contexts in which engineering occurs matter, since cultural contexts can legitimately be said to influence engineering decisions. Engineers need to be aware of relevant cultural norms—in relation to both people in and from other societies—and the consequences of the processes and products of engineering activities. Society makes legitimate demands on how engineers carry out their functions. Conceiving these functions in terms of a single social context vastly oversimplifies matters, since most engineering processes and products exist over more than one society, and each society could have different cultural values. In practice, engineers might thus be required to consider the differential impacts of their actions, which will be further discussed in subsequent chapters.

Second, engineers should carefully distinguish between the moral and nonmoral demands made on them in particular social contexts, since they might easily be misled into believing potentially unethical actions are simply matters of cultural practice. Engineers should be aware that simply because individuals *do* commonly engage in certain actions does mean that they *should* engage in these actions.[95] Additionally, simply because actions are generally accepted does not mean that these actions are right. Engineers should consider their actions from the perspective of principles of ethical global engineering, not only in terms of cultural practices. For example, simply because environmental destruction commonly occurs in a given society—and is widely accepted—does not mean engineers are ethically justified in carrying out or participating in projects in that society:

---

94. Again, regarding Japanese values and society, see Nakane (1970) and Smith (1983). For a fuller account of differences between Japanese and US culture and society, see, for example, Rothbaum, Weisz, Pott, and Miyake (2000), Yamagishi and Yamagishi (1994), Markus and Kitayama (1991), and Lincoln and Kalleberg (1992).

95. This line of reasoning is commonly referred to as the "naturalistic fallacy," concluding that because something *is* the case that it *should* be the case, deriving an ethical *ought* from a factual *is*.

- If two engineers work exclusively in one country—and all the products for which they are responsible are sold only in this country—would a more cross-culturally inclined engineer have an advantage over another less cross-culturally inclined engineer? Why or why not?

7.6   BASIC ETHICAL PRINCIPLES FOR GLOBAL ENGINEERING: RELATED TO CROSS-CULTURAL VALUES

Based on the above discussion, two more principles should be added to the list of basic ethical principles for global engineering. With slight modifications, these can also serve as principles of ethics for organizations. Although both principles are implied in the list of basic principles, explicitly highlighting them as additional responsibilities for engineers is helpful, since the approach taken here deals with international and cross-cultural contexts specifically.

7.6.1   Nonmoral Cultural Values: Engineers Should Endeavor to Understand and Respect the Nonmoral Cultural Values of Those They Encounter in Fulfilling Their Engineering Duties

If engineers fail to develop adequate understandings of practices belonging to cultures other than their own—and they are obliged to communicate with engineers from or the public of these societies, then they will be unable to practice engineering in a competent manner. Engineers cannot simply assume that the practices they have learned—based on a particular set of social values—will also always be effective in societies based on different value systems:

- Give an example of how respecting a culture's nonmoral values could benefit an engineer and/or the company for which he or she works.

7.6.2   Cultural Values and Ethics: Engineers Should Endeavor to Refuse to Participate in Engineering Activities That are Claimed to Reflect Cultural Practices But That Violate Basic Ethical Principles for Global Engineering

Engineers have a duty to follow basic ethical principles for global engineering. Nonmoral values do not override ethical requirements, since ethical requirements are based on an interdiction against seriously harming others. In other words, the avoidance of harm takes precedence over cultural traditions.

Although the priority of moral claims should be unproblematic, it could be argued that engineers have a much stronger obligation than stated above, namely, to actively oppose the violation of ethical principles. When dealing with cultures other than their own, however, this demand on engineers seems unrealistic: in circumstances such as these, generally engineers will not have sufficient understanding and power to ensure that others adhere to ethical requirements.

This is thus an example of the more general approach to engineering ethics taken here, that engineering ethics should be realistic in nature:

- How could engineers "refuse to participate" in immoral projects? What types of systems could be developed to ensure engineers do not involve themselves in projects that could negatively impact the public?

EXERCISE TWO—BID RIGGING IN JAPAN (PART TWO)

Returning to step 6 of the case-study procedure in the first exercise above, note how principles 7 and 8—introduced in this chapter—apply to the ethical issue you decided was the most important. Do these principles conflict with any of the ones you listed before? If so, then prioritize the relevant principles, giving a brief justification for this priority. Once you have done so, complete steps 7–10 of the case-study procedure with regard to the case of bid rigging in Japan, resolving the ethical issue you chose.

7.7   SUMMARY

The case about bid rigging in Japan highlights the relations between cultural values, the government, and law––considering why bid rigging in Japan might or might not fall under the purview of ethics. As engineering occurs in increasingly international and cross-cultural environments, engineers will likely encounter colleagues, customers, and members of the public whose values differ from their own. A failure to recognize this fact can lead to not only technical misunderstandings but also unethical behaviors. In addressing these differences, normative ethical relativism is an unsatisfactory position: the *fact* that people subscribe to different values does not *justify* the worth of these values. Nevertheless, understanding the general nature of and specific distinctions within cultural values is important to ethical engineering: broadly, cultural values can be of a moral or nonmoral character, either having the potential to significantly impact the lives of other or being merely matters of personal and cultural tastes. Engineers should recognize that moral values take precedence over nonmoral values, and simply because particular behaviors are widely accepted and/or practiced does not mean that they are right. Finally, although implied in the basic ethical principles for global engineering, two additional principles are important to keep in mind when working in international and cross-cultural environments, where one is likely to encounter different sets of cultural values.

REVIEW QUESTIONS

1. With reference to the case of bid rigging in Japan, explain two positive consequences and two negative consequences of bid rigging with regard to Japanese cultural values.
2. In terms of safety, how are bid rigging and competition related? How can this practice affect work quality?

3. Explain descriptive and normative ethical relativism, highlighting similarities and differences between the two.
4. For global engineering ethics, why is normative ethical relativism an insufficient position? What could be some of the potential consequences of employing normative ethical relativism in global engineering contexts?
5. Define "values" and explain two reasons why the hierarchical nature of value systems can give rise to problems in ethical decision-making.
6. What is the relationship between values and ethics? Why is it important to understand this relationship when evaluating actions considered ethical within various cultures?
7. Explain the difference between moral and nonmoral values. Give three reasons why engineers should be aware of this distinction in cross-cultural contexts.
8. Explain the two new engineering ethical principles outlined above. Provide examples of each principle being followed, and examples of each principle being violated, within a cross-cultural context.

REFERENCES

Caney, S. (1999). Defending universalism. In I. Mackenzie & S. On'Neil (Eds.), *Reconstituting social criticism* (pp. 19–33). London: Palgrave Macmillan.

Garcia, F., Mendez, D., Ellis, C., & Gautney, C. (2014). Cross-cultural, values and ethics differences and similarities between the US and Asian countries. *Journal of Technology Management in China*, 9, 303–322.

Iyori, H., & Uesugi, A. (1994). *The antimonopoly law and policies of Japan*. New York: Federal Legal Publications.

Lincoln, J., & Kalleberg, A. (1992). *Culture, control and commitment: A study of work organization and work attitudes in the United States and Japan*. Cambridge: Cambridge University Press.

Luegenbiehl, H. (2009). Societal values and nuclear power: A case of conflicting priorities. *Humanities and Technology Review*, 28, 43–84. Fall.

Markus, H., & Kitayama, S. (1991). Culture and the self: Implications for cognition, emotion, and motivation. *Psychological Review*, 98(2), 224.

McMillan, J. (1991). Dango: Japan's price-fixing conspiracies. *Economics and Politics*, 3(3). November.

Nakane, C. (1970). *Japanese society*. Berkeley, CA: University of California Press.

Nell, O. (1975). *Acting on principle: An essay on kantian Ethics*. New York, NY: Columbia University Press.

Rachels, J. (2011). The challenge of cultural relativism. In J. Rachels (Ed.), *The elements of moral philosophy* (pp. 12–24). New York: McGraw-Hill Education.

Rokeach, M. (1973). *The nature of human values*. New York: Free Press.

Rothbaum, F., Weisz, J., Pott, M., & Miyake, K. (2000). Attachment and culture: Security in the United States and Japan. *American Psychologist*, 55(10), 1093.

Shafer-Landau, R. (2003). *Moral realism: A defence*. Oxford: Oxford University Press.

Sigurosson, G. (2014). Ethics and ego: East-west perceptions of morality. *Nordicum-Mediterraneum*, 9(2).

Singer, P. (2011). About ethics. In P. Singer (Ed.), *Practical ethics* (pp. 1–15). Cambridge: Cambridge University Press.

Smith, R. (1983). *Japanese society: Tradition, self, and the social order*. New York, NY: Cambridge University Press.

Yamagishi, T., & Yamagishi, M. (1994). Trust and commitment in the United States and Japan. *Motivation and Emotion*, 18(2), 129–166.

Chapter 8

Autonomy

Chapter Objectives

Having read this chapter, completed the included exercises, and answered the associated questions, readers should be able to

- with reference to the exercise of one's sick mother, explain how the relative importance given to autonomy functions in decision-making processes;
- describe the notion of autonomy in general and its relation to conceptions of being human and politics specifically;
- in contradistinction to these other understandings of autonomy, explain the role that professional autonomy plays in engineering ethics.

EXERCISE ONE—PERSONAL AND PROFESSIONAL AUTONOMY: YOUR SICK MOTHER (PART ONE)

To begin to understand the notion of autonomy, the following exercise asks the reader to consider what he or she would do in the situations described below:

1. Your mother is terminally ill in the hospital. You are her only relative. The doctor's prognosis is that your mother has, perhaps, 6 months to live. She is in acute pain, which can only be partially alleviated through medication. The doctor tells you about your mother's prognosis but not your mother. Which of the following would best characterize your reaction?
   a. You agree with the doctor that this decision not to tell your mother is for the best.
   b. You tell the doctor that not telling your mother is wrong, and try to change his or her mind.
   c. You tell your mother the doctor's prognosis while the doctor is absent.
   d. You do not tell your mother the doctor's prognosis directly but give her a variety of hints regarding the likely outcome of her disease.
   e. You study about the disease to be able to decide what to do next.
2. Your mother finds out about the doctor's prognosis. She wants to take an overdose of sleeping pills, which she asks you to get for her. Which of the following would best characterize your reaction?
   a. Refuse her request.
   b. Try to talk her out of this decision.
   c. Talk to the doctor and follow his or her advice.

Global Engineering Ethics. http://dx.doi.org/10.1016/B978-0-12-811218-2.00008-4

d. Do as she asks.

e. Ignore her request.

3. You manage to talk your mother out of taking the pills. Several months pass, and now she is unable to take care of herself but still able to talk. Your mother tells you her suffering is unbearable, and that she knows her life is close to ending. Your mother requests that you administer a fatal dose of medication. Which of the following would best characterize your reaction?

a. Refuse her request.

b. Try to talk her out of this decision.

c. Talk to the doctor and follow his or her advice.

d. Do as she asks.

e. Ignore her request.

4. Another 2 months pass, and your mother slips into a coma from which she is not expected to recover. Her body functions slow to the point that she is unable to breathe on her own, and she requires the help of life support. Your mother has previously given you instructions about what to do if she is in such a condition: she has repeatedly told you she does not want to be kept alive artificially, and you want to honor her wishes. However, the doctor refuses to disconnect your mother from life support. Which of the following would most likely characterize your actions?

a. Disconnect your mother from life support while no one else is present.

b. Follow what the doctor says.

c. Try to talk the doctor into disconnecting your mother from life support.

d. Take the matter to hospital administrators.

e. Get a lawyer and try to have the courts force the hospital to disconnect your mother from life support.

Responses to these questions indicate one's views on autonomy. As with values in general, views concerning autonomy vary both individually and culturally. Take a few minutes to explain and justify the answers you gave: why wouldn't you make different choices? Why are other options less acceptable or unacceptable? What if the patient was not your mother but either a friend or a stranger? Would this change your answers? Why or why not?

8.1 AUTONOMY IN ENGINEERING

This chapter explores the idea of autonomy, especially in the context of engineering. As was discussed in relation to professionalism and businesses environments, autonomy is connected to ethical engineering: for the sake of public safety, at times, it might be necessary for engineers to exercise autonomy, based on their professional expertise. Additionally, autonomy is often considered central to the Western tradition. Thus, when studying engineering in cross-cultural contexts, it is important for students from other traditions to gain an understanding of autonomy. As with values in general, a failure to understand autonomy can result in misunderstandings regarding decision-making processes. For these reasons, this chapter more closely examines the nature of autonomy and its relation to engineering ethics.

8.2   AUTONOMY AS A CONCEPT

"Autonomy" is closely related to individuality and refers to self-determination and independence from coercion, both internal and external, making decisions for oneself. Autonomy must, therefore, be based on knowledge and rationality—giving reasons for the decisions one makes, based on adequate information. Thus, the notion of autonomy supposes characteristics regarding the nature of being human.

Independence of judgment results from autonomous thought, the value of which has been considered "intrinsic" within the Western philosophical tradition. This means that independent judgments are considered valuable in themselves rather than for the accomplishment of other goals.[96] Whether or not human beings are actually autonomous, ideally they would be. Those incapable of arriving at their own independent judgments would lack autonomy, including very young and old people. For these reasons, the aim of "liberal" education—a general course of study to prepare one to live life—is to liberate or free people, gaining information and skills in reasoning to be able to make autonomous decisions.[97] However, freedom is only one part of autonomy. Autonomy also consists in responsibility.

With the freedom to make one's own decisions comes the responsibility for the decisions one makes. Ethical accountability is, thus, central to the ideal of autonomy. For this reason, autonomy can be experienced both as a burden and liberator. If people are incapable of acting autonomously, then their actions are not truly subject to praise or blame. Only those who make decisions—either for themselves or on behalf of others—can be praised or blamed. For example, children are not typically considered autonomous agents, possessing either the knowledge or abilities to make their own decisions and, for this reason, parents are praised or blamed for the actions of their children. This can also be true for adults. For instance, if one person coerces another, then the one being coerced would not be acting autonomously and, for this reason, would not be subjected to either praise or blame. The one coercing the other would be praised or blamed.

"Paternalism" refers to when one person decides or acts on behalf of another, for the benefit of the one on whose behalf a decision is made or action taken. As was mentioned in Chapter 3, paternalism consists in acting like a parent: one person cares about another but does not think the other is capable of

---

96. For surveys regarding the role of autonomy in Western philosophical thought, see, for example, Christman (2015) and Dryden (2015).
97. Regarding the history of liberal education, see Kimball (1986).

making good decisions or taking correct actions, because of the lack of either knowledge or ability:

- Do you consider autonomy an intrinsic good? Why or why not? Do you think this is a result of your personal upbringing, cultural background, or a combination of both? Explain.
- Give an example of where/when autonomy has or has not been important in your own life. Was there an important life decision you made for yourself or that someone made for you?

8.3 AUTONOMY AND ENGINEERS

Autonomy is especially important because of its close connection with professionalism. Again, as was discussed at length in Chapter 3, professions are occupational groups given an especially high degree of autonomy, since neither their knowledge nor their expertise is easily duplicated. Claims for autonomy have traditionally been based on the expertise professionals have developed through long periods of study and training. For this reason, professionals possess an expert authority, previously ascribed to engineers. This authority is connected with autonomy. Professionals should be able to make decisions independent of hierarchical authority, since individuals occupying positions higher up chains of command might not share the expertise of lower ranking professionals. Hence, professions, professional organizations, and individual professionals are advocates for professional autonomy/independence.

In the tradition of professionalism, this advocacy has been framed primarily in terms of the professional-client relationship, where professionals make decisions for clients based on paternalistic motives: as experts, professionals would be in the best positions to determine which decisions and actions are in the best interests of clients. As a result of this knowledge, professionals would also assume responsibilities associated with these decisions and actions. In establishing the professional-client relationship—where professionals decide and act autonomously—the possibility always exists of paternalistic relations, professionals assuming responsibilities for decisions and actions on behalf of clients. The more autonomy is given to professionals, the less autonomy is available to clients. To ensure that this relationship does not become dominated by the variable—and, perhaps, arbitrary—judgments of individual professionals, professions are expected to exercise functions of control and sanction in relation to individual professionals. This refers to the contract model of professionalism discussed at length in Chapter 3, where professional organizations exercise controls over individual professionals.

Given this framework, it is easy to understand the importance of the use of reason and the acquisition of knowledge, as well as why occupational groups aspire to autonomy for their members. Additionally, this framework makes clear the prestige given to professional status: individuals and groups characterized by professionalism are believed to have the highest amount of autonomy relative to other occupational groups.

Although engineers are given a degree of professional status in societies that emphasize professionalism, as was mentioned before, their status is still somewhat ambiguous. This ambiguity results from the fact that the professional-client model does not neatly fit engineering, since most engineers are first and foremost employees. The closest most engineers have to clients are their employers. Although engineers might work closely with the clients/vendors of the firms for which they work—especially in manufacturing and with regard to supply chains—engineers are ultimately beholden to their employers. Employers are in positions of authority over employees. Clients, by contrast, need help and seek out professionals as a result, such that clients are in positions of natural subservience to professionals. However, even within the framework of professional-client relationships, many have argued for increased client autonomy, based on the claim that all should be able to make decisions for themselves.

Debates surrounding this issue center on the extent to which "correct" answers should count in the decision-making process. Those who argue for client autonomy would hold that the freedom to make decisions for oneself is of higher value than arriving at the best decision. Further, they might argue that professionals are not in positions to understand the broader interests of clients, which encompass more than merely technical concerns.[98] Being able to argue for the importance of expert authority in the case of employed engineers, thus, requires more.

One must show that arriving at the best decisions is more important in the sphere of engineering than in other fields. This can be established by referring back to the first basic ethical principle for global engineering: "Engineers should endeavor, based on their expertise, to keep members of the public safe from serious negative consequences resulting from their development and implementation of technology." Given the importance of individual autonomy in many cultures, to further support this claim, a distinction between the value of autonomy in general and professional autonomy specifically should be kept in mind.

8.4 PERSONAL AND POLITICAL VERSUS PROFESSIONAL AUTONOMY

A clear difference exists between the role an individual occupies as a citizen and participant in the society and the role an individual occupies as a professional. The justification for autonomy in the role of citizen and social participant might occur at two levels. The first consists in the essentialist claim that individuals are, by their

---

98. This position can be seen, for example, in the medical field, where the patient's concerns supersede the concerns of healthcare professionals. For more on the divergence of the priorities of patients from healthcare professionals, see Stronks, Strijbis, Wendte, and Gunning-Schepers (1997).

very nature as rational beings, autonomous and deserving of autonomy. The second consists in the instrumentalist claim that individuals should be allowed to exercise autonomy for the sake of the proper functioning of society, based on ideals associated with liberal democracy—in other words, the claim society works the best when individuals decide and act autonomously. Neither of these two justifications applies directly to claims regarding professional autonomy.

In fact, with regard to autonomy, many generally argue that the most important potential conflict would be between professional duties and personal beliefs—in other words, duties following from and associated with a professional role and the basic moral beliefs of individuals—which could result in, for example, refusing to follow the instructions of a superior. While it would lead beyond the scope of topics under consideration here, it should be noted that manners of resolving such conflicts form one of the more contentious elements of debates within the field of professional ethics.[99] Clear from such debates, however, is this potential conflict between professional and personal ethics, resulting from the fact that the respective justifications for these two sets of ethics are different. Similarly, as a role ethics, justifications for engineering ethics are ultimately based on ensuring public safety.

Claims of autonomy for engineers should, therefore, be justified in terms of their reasons for existing within society: engineers exist to design, develop, and implement technologies, and, within these contexts, engineering ethics should ensure public safety. Claims of autonomy for engineers are, thus, justified with reference to the necessity of autonomy for engineers to properly carry out tasks assigned to them by society, rather than with reference to an essentialist or instrumentalist ideal of autonomy as central to either being human or social functioning.[100] Assuming this line of reasoning, engineers would be justified, for instance, in refusing to follow the orders of superiors in circumstances where following these orders would endanger public safety. In reaching such a conclusion, however, it is important to keep in mind other previously discussed duties that engineers have, duties related to their roles as employees, business ethics, and the recognition of cross-cultural concerns. The next chapter further considers the potential for conflicts among these demands, and explores possible solutions to these conflicts:

- Give an example of a conflict between professional duties and personal ethics and how this conflict could be resolved.
- Give an example of a situation in which an engineer might have to exercise professional autonomy in relation to a superior who does not have the necessary expertise to make the best decision.

---

99. While this debate is not heavily featured in discussions of engineering ethics, it can be seen in many subdomains of professional ethics. For example, see Rassin (2008), Singhapakdi and Vitell (1993), Eastman, Eastman, and Eastman (1996), and Hazard (1992).
100. For a fuller consideration of the role of autonomy within engineering—in relation to professionalism, public safety, and cultural values—see Luegenbiehl (2007).

EXERCISE TWO — PERSONAL AND PROFESSIONAL AUTONOMY (PART TWO)

Returning to your answers from the exercise at the beginning of this chapter, explain where and how your answers indicate the relative priority you give to either personal or professional autonomy.

8.5 SUMMARY

As the exercise concerning one's sick mother makes clear, to a large extent, the importance attached to autonomy can vary both individually and culturally. Nevertheless, as was discussed in previous chapters on professionalism and business ethics, the notion of autonomy is central to engineering ethics. Autonomy consists in not only making decisions and taking actions for oneself and others, but also taking responsibility for the consequences of these decisions and actions. Rather than framing autonomy in fundamentally essentialist or politically instrumentalist terms—where either individuals are conceived as intrinsically autonomous or autonomous actions are supposed to serve democratic social ends—autonomy for engineers is based on the role responsibilities of engineers. Insofar as engineers are professionals with obligations to public safety, engineers are justified in acting autonomously and, at times, paternalistically.

REVIEW QUESTIONS

1. Describe the nature of personal autonomy and two situations where one would not be acting autonomously.
2. Explain "liberal" education and its relation to the ideal of autonomy.
3. Describe the nature of paternalism and explain its relation to autonomy and engineering ethics.
4. With regard to autonomy in the field of engineering, discuss problems associated with the professional-client model.
5. Explain two reasons autonomy would be justified for citizens and social participants. How are these different from the justification of autonomy for engineers?
6. When confronted with a directive by a supervisor who lacks professional expertise in the field of engineering, how might one exercise professional autonomy?

REFERENCES

Christman, J. (2015). Autonomy in moral and political philosophy. In E. Zalta (Ed.), *The Stanford encyclopedia of philosophy.* http://plato.stanford.edu/archives/spr2015/entries/autonomy-moral/.

Dryden, J. (2015). Autonomy. *Internet Encyclopedia of Philosophy.* http://www.iep.utm.edu/autonomy/.

Eastman, K., Eastman, J., & Eastman, A. (1996). The ethics of insurance professionals: Comparison of personal values versus professional ethics. *Journal of Buisness Ethics, 15*(9), 951–962.

Hazard, G. (1992). Personal values and professional ethics. *Cleveland State Law Review, 40,* 133.

Kimball, B. (1986). *Orators & philosophers: A history of the idea of liberal education.* New York, NY: Teachers College Press.

Luegenbiehl, H. (2007). Ethical autonomy and engineering in a cross-cultural context. *Techné: Research in Philosophy and Technology, 11*(1). https://scholar.lib.vt.edu/ejournals/SPT/v8n1/luegenbiehl.html#luegenbiehl2003 Fall.

Rassin, M. (2008). Nurses' professional and personal values. *Nursing Ethics, 15*(5), 614–630.

Singhapakdi, A., & Vitell, S. (1993). Personal and professional values underlying the ethical judgments of marketers. *Journal of Buisness Ethics, 12*(7), 525–533.

Stronks, K., Strijbis, A., Wendte, J., & Gunning-Schepers, L. (1997). Who should decide? Qualitative analysis of panel data from public, patients, healthcare professionals, and insurers on priorities in healthcare. *British Medical Journal, 315*(7100), 92–96.

Chapter 9

Conflicting Duties and Dissent

Chapter Objectives

Having read this chapter, completed the included exercises, and answered the associated questions, readers should be able to
- explain the nature of loyalty and of legitimate authority and their relation to faithful agency;
- describe the nature of and give specific examples of conflicts of interests;
- explain the different forms of whistle-blowing and conditions under which whistle-blowing would be morally permissible and required.

CASE STUDY ONE—THE CASE OF JOHN'S FRIENDSHIP

John is a design engineer working for a major car company, the Carbon Car Company (CCC). He has been working on the development of a new sports car. One day at work, his wife, Audrey, calls saying his best friend since childhood, Ken, has been killed in an automobile accident. John attends the funeral, at which time he discovers Ken had been driving a CCC vehicle, a Rebel SUV (Sports Utility Vehicle). John's vehicle rolled over while traveling approximately 60 mi/h on a straight road. John commiserates with Ken's widow, Mary, and three children, all of whom he has become very fond over the years. In fact, John introduced Ken to Mary.

After returning home, John throws himself back into design work, generally trying to forget about the tragic accident. Every once in a while, however, he hears about performance problems associated with the Rebel SUV. John generally ignores these, since he is busy with his own design work on a sports car.

Approximately 6 months later, John receives a distraught call from Mary, telling him CCC has disclaimed all responsibility related to Ken's accident, claiming it was solely the result of driver error. Since John works for CCC and knows a great deal about cars, Mary wants to know if there is anything he can do to help her with the case. John decides to investigate the issue by examining a number of confidential company documents on his work computer. To John's surprise, he discovers a significant number of reports related to Rebel SUV rollovers—more than a few of which have resulted in fatalities. Comparing internal company statistics with publicly available data, John realizes the Rebel has approximately five times the industry average of rollovers. He does not recall ever seeing any public reports regarding this information.

Global Engineering Ethics. http://dx.doi.org/10.1016/B978-0-12-811218-2.00009-6

EXERCISE ONE—THE CASE OF JOHN'S FRIENDSHIP (PART ONE)

With regard to the case of John's friendship, complete steps 1–10 of the case-study procedure, using all relevant principles reviewed thus far.

9.1   AUTONOMY AND DISSENT

The previous chapter discussed the notion of autonomy for engineers, in terms of the claim engineers should be able to exercise independent judgment based on their professional expertise: if engineers have a responsibility to ensure public safety while introducing new technologies into society, then they should have the freedom to do so. Otherwise, engineers should not be held responsible for their actions. As mentioned previously, however, that perspective is based on engineering alone. This chapter further examines some of the potential limits on the exercise of independent judgment by engineers, based on other roles they occupy.

9.2   THE DUTY OF LOYALTY: A SPECIAL BOND OF IDENTIFICATION

Engineers have a duty to the public, a claim established in previous chapters. Likewise, however, they are also employees and, as such, they have a duty of obedience, stated in the first principle of employee ethics: "Corporate employees should endeavor to obey all legitimate, job-related directives." This duty can be conceived in terms of the broader context of loyalty.

Codes of engineering ethics have traditionally emphasized loyalty in a narrower sense—in terms of a duty of loyalty to an employer or client, providing the basis for discussions of conflicts of loyalty. From the perspective of global engineering, however, the issue is both more fundamental and complex. In a sense, it could be considered the most fundamental problem in global engineering ethics: in many societies, the issue of loyalty to something other than engineering seemingly overrides all other considerations. Hence, making loyalty the basis for ethical discussions potentially creates real problems.

"Loyalty" refers to a special bond established between two parties, individuals, group of persons, or institutions. Loyalty could even be ascribed to an ideal, such as social justice. This means that the relationship is a close and enduring one, involves trust between the parties, and that a relationship of identification exists with the object of loyalty, a relation of care and concern. Discussions often center on loyalty to one's country, family, social group, or even to pets. One's occupation can also be an object of loyalty, as can one's employer.

The importance of loyalty here consists in the fact that, when one becomes an engineer, one has a duty of loyalty to engineering as a profession and that, when one becomes an employee, one has a duty of loyalty to the company for which one works. More specifically, one should care and be concerned with the interests of engineering and one's employer. As was discussed previously, in becoming an engineer and receiving a paycheck, one owes a special allegiance to the professional group and company. Of course, all people have multiple loyalties.

Professors, for example, have loyalties to their families, universities, professional organizations, countries, and so on. Generally, they are capable of placing these loyalties hierarchically, based on their values. A particular professor, for instance, might have a greater sense of loyalty to her university than to the professional organizations of which she is a member, such that if these two parties made competing and mutually exclusive demands on her, then those associated with her university would take priority. As with role responsibilities and values in general, conflicting loyalties can, thus, usually be resolved. A problem occurs when the demands for loyalty by two parties are absolute and they conflict. In a situation such as this, the professor would have two absolute duties and be unable to fulfill both of them. Conflicting loyalties can, therefore, create moral dilemmas for engineers, which they have to resolve in the absence of preexisting priorities:

- To whom should engineers have greater loyalty, the public or their employers? Justify your answer.

9.3   THE LEGITIMATE AUTHORITY OF EMPLOYERS

As mentioned before, the roles of employers are different from those of clients: clients hire engineers to complete very specific tasks, and engineers generally decide how to carry out these tasks, receiving agreed-upon fees for the completion of these tasks. By contrast, employers hire engineers for generalized roles, where specific assignments arise throughout employment rather than only prior to it, receiving salaries. If someone is hired as a mechanical engineer by a company, for example, that person can be assigned to work exclusively as a mechanical engineer or on a variety of engineering projects, but not to make coffee. The authority of an employer over an engineer is, thus, based on the conditions in terms of which the engineer was hired. These conditions are often established through an employment contract.

In contracts, the duties of an employee to a company are outlined, as well as the wages the company will pay the employee for carrying out these duties. To be fair, the contract should be established in an agreement between equals. In reality, however, this is rarely the case. To a large extent, agreed-upon conditions depend on alternative employment opportunities, employment location, financial pressures, and so on. More often than not, companies will be in positions of greater power than potential employees and, thus, able to dictate coercive employment conditions. This can be especially true with younger engineers or those just beginning their careers. The notion of legitimate authority, therefore, becomes important in cases such as these: those in positions of authority are able to give orders, but engineers are not always obliged to follow all such orders.

As mentioned previously, exercising legitimate authority consists in giving orders to subordinates that would be considered "legitimate," the basis of which would be the contract to which the employee freely agreed. Further, companies cannot engage in unethical activities. Signing an employment contract to work for a criminal organization, for instance, illegitimizes directives the boss of such an organization would give. Even when employees have signed employment contracts, however, the legitimate authority of employers is still limited. Companies cannot violate the basic rights of employees, for example, the right not to be discriminated against. These need not—and generally will not—be stated in employment contracts. Again, employment contracts cannot require employees to do anything unethical, and it can be argued that demands by employers cannot be unreasonable.

Engineering employees are in special situations, since they are professionals. Employers possess authority based on their institutional positions, while engineers possess authority based on their expertise. Again, this means that engineers are required to exercise independent judgment.

9.4   FAITHFUL AGENCY

If companies exercise only legitimate authority, then, in turn, employees have duties to companies; they are compensated for doing what they agreed to. Duties of employees, therefore, consist in following through on the legitimate directives of employers. Performing assigned tasks in an "acceptable fashion" or "well" is related to this broader duty. Again, the criteria of acceptability vary from situation to situation, and specifying conditions under which tasks would be completed well can be difficult. One might claim that a job done well is simply one that meets the expectations of an employer, but these expectations could be unreasonable. Perhaps a more fruitful approach would be one that characterizes the motivations that should govern the performance of employees.

While employed by a particular company, the primary duty of an employee is to that employer. In essence, employers purchase the time and skills of employees. Thus, employees have obligations to contribute to the goals of the companies for which they work, putting these goals before those of mere self-interests. This relation is known as one of "faithful agency." Employees have obligations to act as faithful agents of the companies employing them: in working for employers, employees should be devoted to the needs and interests of the employers. If employees do so, then they will generally be performing their jobs well. (Unless, of course, they are incompetent in other regards—as discussed in Chapter 4, competence is a requirement of professional engineers.) Loyal engineers can, thus, usually be considered to do their jobs well—in other words, loyalty is a condition of a job well done. Again, however, compared with other employees, engineers are in somewhat special positions.

Engineers also have a duty of loyalty to the public, insofar as they are special protectors of public safety, health, and welfare. Thus, for engineers, situations of conflicting loyalties can arise, where loyalties to employers and loyalties to the public conflict. Some have argued that this would not be a real conflict: in protecting the public from potential harms by companies, engineers protect companies from potential liabilities.[101] Consequently, in protecting the public from companies, engineers are also acting loyally to the interests of their companies. Others have argued that loyalty to the public should always come before loyalty to companies, as was discussed in Chapter 6. Still others claim it is the other way around.

In his provocatively titled "I'm Not That Hungry—Yet!," Richard Mayer, an engineer, argues that the duties engineers have to their employers should override their loyalties to the public: "An engineer employed in industry may give the appearance of professionalism, such as authority and responsibility for failure, but these are internal to the company structure. He has no real relationship with the external customer or the public, nor actual responsibility for the outcome. He is not ultimately responsible, for he is not free to exercise independent judgment. He is an employee, paid by the company, and he is responsible to it alone. It is the company's reputation and profit that are at stake, not his own" (Mayer, 1977).

Mayer goes on to frame this claim in terms of faithful agency: "The proper role of the engineer in industry is that of the 'good and faithful servant.' He can bring to the assignment his best skills, but not professional independence, for responsibility cannot rest both with the engineer and with the company. This, of course, was the engineer's own choice—to trade his skills for security. While others served their additional apprenticeship to the public, he was drawing a paycheck. He did not enter practice; he got a job. We need both employed and professional engineers, but one cannot have it both ways" (Mayer, 1977).

- As outlined in the text, some argue that there is not a conflict of interests between loyalty to a company and loyalty to the public, that "in protecting the public from potential harms by companies, engineers protect companies from potential liabilities." Do you agree with this claim? Why or why not?

9.5   CONFLICTS OF INTERESTS

In framing his claim for faithful agency, Mayer points toward a duty engineers have to avoid conflicts of interests: acting autonomously, based on their duties to the public, can place engineers in situations of conflicting interests with their

---

101. For an example of this position, see Vandekerckhove and Commers (2004).

employers. Conflicts of interests can occur when the judgments or actions of engineers on behalf of employers or clients are endangered by other interests or commitments, for instance, those to family, country, themselves, alternative potential clients or employers, or their professional roles. When engineers allow these to override their primary duties to employers, one could claim that engineers are disloyal. Loyal employees should avoid conflicts of interests.

Philosopher Michael Davis has characterized three different types of conflicts of interests: (1) actual—when the interests of employers are certain to be negatively affected, for example, an employee selling proprietary information to a competitor; (2) latent—when there is a reasonable chance that the interests of employers will be negatively affected, for example, an employee working under the direct supervision of a close friend or family member; and (3) potential—when one can reasonably foresee that the interests of employers might be negatively affected, for example, an employee taking an intensive, second job. Some of these conflicts, Davis notes, cannot always be avoided. As a solution, he proposes full disclosure to employers and consent by employers to the conflicts.[102]

Examples of situations where conflicts of interests can occur include, but are not limited to, when engineers have second, part-time jobs, use information learned at their employers for personal gain, are offered gifts or bribes, or show favoritism in hiring/subcontracting work. None of these situations require that actual conflicts of interests occur, although they are all sources of possible conflicts. Additionally, from an outside perspective, it is not easy to determine when conflicts of interests are occurring.

In the case of gifts, for example, the following questions would need to be considered: how much is it worth? Why is the gift being given? How is it presented? What are the positions of the giver and recipient of the gift, respectively? What are the company policies? What are the common practices within particular industries regarding giving and receiving gifts? Receiving a gift worth $50 might bias the judgment of one engineer, for instance, whereas $5,000 might bias the judgment of another. This is one of the reasons that conflicts of interests are the sources of such intense ethical debates. However, the distinction between actual and apparent conflicts of interests somewhat simplifies matters: even if the judgments of engineers are not negatively influenced, to an outsider, they might appear to be. From the perspective of corporations, this situation can be just as bad as a real conflict. One might thus argue that engineers should avoid situations not only that result in *actual* conflicts of interests but also that have the potential to result in *apparent* conflicts of interests.

Engineers need to handle dealing with both actual and potential conflicts of loyalties, duties of loyalty to employers and the public. This raises the question of dissent within corporate environments—if, when, and how to disobey orders.

---

102. For more on this, see Davis (1998).

CASE STUDY TWO—THE CASE OF LARRY SAPPORO: CONFLICTS OF INTEREST?

Read through each of the following hypothetical scenarios. Take a few minutes to respond to the issues posed after each section, mentioning any potentially relevant ethical principles to justify your answers.

Larry Sapporo has been working as a chemical engineer for Plesticine Corporation for the last 5 years, since he graduated from university. When Larry began working for the company, he signed an industry standard nondisclosure agreement, agreeing not to reveal proprietary company information. Although Larry has contributed significantly to the company, his career with Plesticine seems to have reached a plateau. He has not been promoted to a managerial position since he joined the company. Larry is ambitious and hopes, someday, to move into a leadership position at a major company. To move toward this goal, Larry believes he should exhibit leadership qualities, although, unfortunately, such opportunities at Plesticine have been few and far between. One day, Larry receives a call from a representative at Lysotane Corporation, a company in the same industry as Plesticine. The representative indicates that he has heard rumors that Larry might be interested in changing jobs. Lysotane currently has a job opening for someone like Larry, one that would consist in leading a development team for new products. The position would pay approximately twice as much as Larry's current salary, and his benefits would be negotiable:

1. Should Larry speak with his current employer before engaging in further discussions with Lysotane?
2. Should he be suspicious of job offers from companies competing with Plesticine?
3. Should Lysotane refrain from attempting to hire away engineers from their competitors?

Larry decides to accept the position and begins working at Lysotane. In his first assignment, Larry realizes that one of the chemical processes he used at Plesticine could be used to develop a new product for Lysotane. This product would not compete with those produced by Plesticine. Larry learned of the theory behind this process while studying at university, although never heard of its being used anywhere except Plesticine. Larry does not know whether Plesticine would consider this process to be proprietary information:

1. Should Larry use this process at Lysotane, since he learned the theory behind it in college?
2. Is his knowledge of this process part of a general engineering knowledge constituting his experience base?
3. Should Larry use this process, since the product for which it would be used to produce would not compete with ones produced by Plesticine?
4. Should he consult Plesticine, to see whether the company considers this process proprietary?

Larry incorporates the process into the design of the new product, and Lysotane makes plans for production. Having completed this task in his first 6 months, Larry is assigned to lead another team in the development of a different product. Plesticine's most successful product is Garbonzo. His superiors tell Larry that Lysotane needs a product to compete with it. When he was at Plesticine, Larry worked on the development of Garbonzo. Since Garbonzo was already a "mature product," this was part of the reason that he was unsure of his future at Plesticine—there was little room for innovative development:

1. Should Larry agree to work on the development of a product that directly competes with Garbonzo?
2. Should he consult Plesticine before doing so, seeking out information regarding the parameters of proprietary information regarding the knowledge he would use to develop a product that would compete with Garbonzo?

Larry accepts the assignment, but work comes along very slowly. He cannot develop an alternative product with the same quality as Garbonzo. At this point, Larry is tempted to use a process he learned at Plesticine, although he knows that Plesticine has explicitly designated this process a trade secret:

1. To which organization does Larry owe greater loyalty, Lysotane or Plesticine?
2. If he can ensure that Plesticine will not discover his actions, then should Larry use the process he learned there to develop a product for Lysotane to compete with Garbonzo?
3. Do his obligations to Plesticine cease once it stops paying him?

Larry decides not to use the process, although his superiors at Lysotane keep pressing him for results. Given how much he learned during his time at Plesticine, his superiors hint to Larry that he should really be able to do more to develop a competitive product. Finally, they directly tell him that the only reason he was hired by Lysotane is because of the knowledge he acquired at Plesticine. His superiors tell Larry that if he fails to use this knowledge to develop a product that can compete with Garbonzo, then he will lose his job at Lysotane.

Unfortunately, since Larry moved to Lysotane, the job market for chemical engineers has become quite bad, and he has no prospects of alternative employment. To make matters worse, one of his children requires major medical work, which Larry can only afford to pay with his employee insurance plan:

1. To what extent is Larry's current predicament the result of his earlier decision to take the position at Lysotane?
2. Should Larry give up his job at Lysotane?
3. Does he have an obligation to do whatever is necessary to ensure his child's medical problems are handled?
4. Should Larry tell Plesticine what Lysotane is attempting to do?

9.6   ENGINEERS AND DISSENT

Based on previous considerations, engineers might have to decide if and how to dissent from hierarchical directives—to circumvent corporate chains of command. In relation to cultural practices, Chapter 7 addressed one form of dissent: "Engineers should endeavor to refuse to participate in engineering activities which are claimed to reflect cultural practices but which violate the general ethical principles of engineering." The application of this principle is limited to situations in which ethical considerations compete with nonethical ones, however, thereby relying on the more fundamental status of ethics—in terms of behaviors that have the potential to seriously impact the lives of others. Situations of dissent being considered here are more problematic, since they touch on competing ethical claims—duties of loyalty to the public versus employers.

Dissent in corporate environments can take different forms, consisting in any one of the following: disagreeing with a superior, either orally or in writing; refusing to carry out the directives of a superior; going above a superior, to a superior's superior; or revealing internal information to external individuals and/or organizations. The last of these is generally known as "whistle-blowing" and is the most serious. For this reason, it will serve as a model for dissent: important considerations with regard to whistle-blowing also apply to other less serious forms of dissent, to lesser degrees. Given the seriousness of whistle-blowing, it should be mentioned that the majority of engineering ethical issues will not require contemplating dissent within corporate hierarchies.

9.7   ENGINEERS AND WHISTLE-BLOWING

Whistle-blowing is a complex and disputed topic, with concerns revolving around its nature and consequences, organizational structures and ways to minimize the need for whistle-blowing, requirements for its implementation, motives of whistle-blowers, and so on. Whistle-blowing always involves conflicts of loyalties, which the philosopher Sissela Bok describes as follows: "loyalty to colleagues and to clients comes to be pitted against loyalty to the public interest, to those who may be injured unless the revelation is made. Because the whistle-blower is an insider in the very organization he criticizes, his act differs from muckraking and other forms of exposure by outsiders, as when reporters expose corruption within a governmental agency. Such acts are expected, sometimes even required of outsiders, and do not produce in them the same conflicts of loyalty" (Bok 1980).

To begin, it is necessary to articulate a definition of whistle-blowing suitable for engineering. It can be defined as follows: *the violation of a corporate hierarchy to make known a clear danger or problem in an organization that is likely to cause serious physical harm to the public.* Based on this definition, it is possible to distinguish between two forms of whistle-blowing, internal and external.

Internal whistle-blowing occurs when a hierarchical chain of command within an organization is broken, such that immediate supervisors are bypassed, perhaps because they have refused to act or have themselves been involved in wrongdoing. External whistle-blowing occurs when information internal to an organization is shared externally, perhaps with regulatory agencies, the press, or public. The philosophical literature often restricts the use of the term to external whistle-blowing, although the media typically uses the term in both senses. A further distinction can be made between open and anonymous whistle-blowing.

In the first form, whistle-blowers reveal their identities, while in the second, they remain unnamed. Studies have shown that anonymous whistle-blowing is less effective, since it is easier to ignore, and no follow-up with the whistle-blower is possible. Organizations have also been willing to devote significant resources to discovering the identities of anonymous whistle-blowers. This is relatively easy since only a limited number of individuals would typically have access to the kinds of information revealed. Whistle-blowers often want to hide their identities, since they almost always suffer due to reprisals from organizations or colleagues.

Finally, it should be mentioned that whistle-blowers need to be insiders—either currently or formerly associated with the organizations on which they blow the whistle. Otherwise, they might be spies, reporters, or moles, but not whistle-blowers. As mentioned above, this is due to the fact that whistle-blowing involves conflicts of loyalties—between duties of loyalty to organizations and those to the public or other principles. For those who infiltrate organizations, no such duties of loyalty to these organizations exist:

- Whistle-blowing is defined here as "the violation of a corporate hierarchy to make known a clear danger or problem in an organization that is likely to cause serious physical harm to the public." In your opinion, what constitutes serious physical harm? Is whistle-blowing justified when the harm is financial or emotional rather than physical? Why or why not?

9.7.1 Criteria for Whistle-Blowing

The philosopher Richard DeGeorge is responsible for one of the most influential discussions of whistle-blowing. He outlines five conditions according to which whistle-blowing would be either morally permissible or morally required. Reviewing these criteria allows for further insights into the nature of whistle-blowing. The first three conditions establish the permissibility of whistle-blowing, and the last two its necessity. From a moral point of view, whistle-blowing is permitted if

1. serious harm will be done to the public;
2. the employee makes his or her superiors aware of the problem;
3. the superior does nothing and the employee exhausts the corporate hierarchy.

If an engineer blows the whistle after these steps have been unsuccessful, at some cost, then whistle-blowing is morally permissible. According to DeGeorge, an act such as this deserves praise, but a failure to blow the whistle in a situation such as this would not deserve moral blame. However, whistle-blowing becomes morally obligatory if two additional conditions are met:

4. The whistle-blower has documentary evidence available, which would convince an objective observer.
5. It is very likely that, as a result of the whistle-blowing, the problem would be solved.

The primary focus of DeGeorge's account of whistle-blowing is on the effectiveness of whistle-blowers—how effective it is in keeping the public safe. He does not consider the consequences to whistle-blowers to be a significant factor—how whistle-blowers will be affected by their actions. DeGeorge recognizes that although engineers may become public heroes for whistle-blowing, their colleagues will most likely consider them traitors—perhaps because whistle-blowers serve as reminders of their own moral failings. In general, the literature is full of examples of the fates of whistle-blowers. Although willing to list whistle-blowing as a potential requirement, professional societies have not always stood behind whistle-blowers. This is irrelevant, according to DeGeorge, if the potential harm is great enough.[103]

Due to the serious consequences associated with whistle-blowing, however, most stress that it should be an avenue of last resort.[104] Many recent discussions have emphasized ways to avoid whistle-blowing, such as setting up ethics offices within organizations, fostering open-door practices, articulating clear organizational policies, and appointing ombudspersons—officials who mediate employee grievances with management.[105] These alternatives are emphasized because whistle-blowers often become the targets of subsequent investigations, rather than the misconduct revealed by the whistle-blowing.

Within organizations, the consequences for whistle-blowers are so universally negative—including shunning by colleagues and formal and informal organizational reprisals—that whistle-blowing can legitimately be considered an act of moral heroism. As such, it is unclear that whistle-blowing should be considered a moral requirement, as has been the case with engineers in particular. While technical professionals often possess more information than others, the possession of this information alone does not establish a duty to blow the whistle. Moral heroes should be applauded and rewarded, as occurs through various governmental incentive schemes.[106] However, singling out individuals—demanding they sacrifice their careers and suffer other detrimental consequences for the common good, to avoid charges of moral misconduct—is debatable.[107]

103. For a fuller account of DeGeorge's position with regard to whistle-blowing, see DeGeorge (1999).
104. Regarding the consequences of whistle-blowing, see Lubalen and Matheson (1999).
105. For examples of these discussions, see Dandekar (1991), Davis (1989), and Gunsalus (1998).
106. For example, see Callahan and Dworkin (1992) and Castellina (2011). For a comparison of these schemes and their European counterparts, see Fleischer and Schmolke (2012). For more on European whistle-blowing schemes, see also Huttl and Lederer (2013).
107. For a fuller discussion of whistle-blowing in general, see Luegenbiehl (2005) and Martin and Schinzinger (2010) and discussions of particular issues related to whistle-blowing in Davis (1983) and Martin (1992).

The issue then is whether whistle-blowing is an act that can be ethically required of engineers under certain circumstances. As with the overall approach here, this is an issue that should be resolved with regard to individual cases. The following is an overview of reasons given in opposition to and favor of whistle-blowing:

In Opposition to Whistle-Blowing

- It is the action of a traitor who will hurt organizations and their employees.
- Whistle-blowers position themselves as morally superior to those who did not act.
- Engineers owe their greatest loyalties to the organizations for which they work.
- Whistle-blowing is often motivated by reasons that have little to do with public safety.
- Whistle-blowers could be mistaken, thus causing unnecessary harm to organizations.
- Whistle-blowers could destroy their careers.
- Engineering organizations might not support whistle-blowers.
- Engineers should not be held responsible for the actions of others.

In Favor of Whistle-Blowing

- The greatest professional duty of engineers is to the public and its safety.
- Great harm can be done if such wrongs are not corrected.
- Engineers have a right to free speech.
- The public will admire engineers as heroes.
- As members of organizations, engineers share in the responsibility of actions taken by those organizations.
- Engineers have the greatest amount of technical knowledge and, therefore, can make the most convincing case to the public.

- Should engineers be held responsible for the actions of others within their organizations? Why or why not?
- Do you agree that whistle-blowing should be considered an act of "moral heroism"? Why or why not?

CASE STUDY THREE—THE CASE OF GEORGE KIRIN

Read through each of the following hypothetical scenarios. Take a few minutes to respond to the issues posed after each section. In answering these questions, refer to ethical principles discussed in previous chapters and criteria for whistle-blowing reviewed in this chapter to justify your answers.

George Kirin works as a design engineer for a second-tier automobile manufacturer, the Multiple Automobile Assembly Corporation (MAAC). He is an experienced engineer, 12 years out of university, who has worked in a variety of design positions for three different companies in the industry. His work has been highly regarded by all of his employers.

George is currently working with a team responsible for the design of the rear-end brake assembly of a new, highly fuel-efficient model that MAAC hopes to introduce in less than two years. Code-named "Springboard," the company is pinning its hopes of "leaping" from a second- to first-tier automobile manufacturer on this model. If it fails, then MAAC will be relegated to hanger-on status, perhaps even bankruptcy.

If nothing else, George is an extremely conscientious engineer. He has been performing extra tests outside the scope of his assignment and has discovered what he thinks is a flaw in the overall rear-end design of the car. Under certain conditions, this flaw will cause the brakes to malfunction:

1. Should George be undertaking design tasks aside from those assigned to him?
2. Is he acting in a manner similar to most responsible engineers?

George decides to inform his superior of the problem. His supervisor, Colleen Keystone, tells him that the issue is outside the scope of the design assignment of their department. When George reiterates his concerns, Colleen tells him she will review his data.

Several weeks later, after repeated inquiries by George, Colleen calls him into her office for a meeting. She explains to him that her examination of his data indicates that the design is acceptable at present. In talking with her, George is unconvinced Colleen has carefully examined the problem in relation to the overall rear-end assembly design:

1. Is George exhibiting sufficient respect for Colleen's judgment?
2. Is her perspective on the responsibilities of design engineers too narrow?
3. At this point, should either George or Colleen alert upper-level management of the problem?
4. Are there problems with the design process of MAAC?

After a greater deal of reflection—and additional tests that seem to verify his initial concerns—George decides to send a memo to the head of the whole design team, Sammy Adams, listing and justifying his concerns. He does not receive a memo in reply, but approximately a month later, George receives a call from Sammy telling him to drop the matter.

When George presses the issue, he is told that while a potential problem does exist, fixing it at this stage would delay the project, to the point that the model of another major company would beat theirs to market. This delay would prevent MAAC from gaining sufficient market shares with their new model, killing the company. If the project goes ahead as planned, however, then the company would be able to address any potential liability claims with profits from the new model. Sammy tells George that he needs to understand the business realities facing their company and that MAAC has considered all available options:

1. In your opinion, has George followed appropriate procedures in communicating his concerns? Why or why not?
2. Is Sammy trying to cover up the problem?
3. Should MAAC be primarily concerned about its survival?

At this point, George asks for permission to examine the matter further, forming a small team to look into additional options. He is given permission to do so. However, the only option George and his team discover, which would fit into the current timetable, would significantly lower the fuel efficiency of the car. When he reports these results, George is told that this is an unacceptable option:

1. Is the fuel efficiency of a car as important as its brake system?
2. Should George cease work on the design of the brakes?
3. If MAAC refuses to change the design, then should George make his concerns public?

The car goes into production on schedule and is extremely successful. Very soon, however, the company becomes aware of isolated instances of malfunctioning brakes, which are handled through warranty procedures. During the first year, out of the approximately 260,000 units sold, 130 cases of brake malfunctioning are reported. In the meantime, George has been working on another project and is unaware of these statistics.

Finally, just over a year after the car was introduced, a sensational case occurs: the loss of brakes is the apparent cause of one of the cars smashing into twelve elementary school children, killing eight of them. George first hears about the case on the news and then follows the story closely. Litigation against MAAC begins with lawsuits in excess of $800 million. MAAC takes the defensive, claiming the brake failure was an anomaly isolated to that particular car, neither a design nor a manufacturing problem. The company blames the driver for the collision.

In the meantime, George has obtained failure data on other cases from Steve Johnson, a friend working in another division of MAAC. Steve has shared this data with George, despite a company policy prohibiting the interdivisional sharing of information without proper authorization:

1. Should MAAC have issued a recall on the brakes before the school children incident?
2. Should Steve have given the data to George?
3. Is either MAAC or George responsible for the deaths of the school children?
4. Should George share the data he has obtained with the plaintiffs suing MAAC?
5. Should George contact the media and reveal what he knows about the design procedure?

EXERCISE TWO—THE CASE OF JOHN'S FRIENDSHIP (PART TWO)

Refer again to the case of John's friendship at the beginning of this chapter. Answer the following questions, where relevant, mentioning the criteria for whistle-blowing to justify the answers you give:

1. Should John's friendship affect his actions? Why or why not?
2. Should his employment with CCC affect John's decisions? Why or why not?

3. What actions should he take, if any? Which actions would be wrong for John to take?
4. What if he takes actions, keeping the matter within the company, and they are ineffective. What should he do?

9.8 SUMMARY

Given the many duties of engineers—based on their professional expertise and the roles they occupy both personally and professionally—the fact that these potentially conflict is unsurprising. This chapter more fully examined the nature of these conflicts and how to avoid or resolve them. If engineers have a responsibility to exercise professional autonomy for the sake of public safety, then they should be allowed to do so. At times, however, this can put engineers at odds with hierarchical authority. Engineers have a duty of loyalty to the organizations in which they work, assuming organizational directives are based on legitimate authority. Engineers should act as faithful agents of the companies for which they work: doing a good job as an employee consists in a relation of faithful agency, taking on the interests of the organization as one's own. Such a relation involves avoiding apparent, potential, and actual conflicts of interests. At times, however, engineers might find themselves in situations where they should dissent from hierarchical authority and directives. The most serious form of such dissent is "whistle-blowing," which can occur in a number of ways. Given the negative consequences for both organizations and engineers that can result from whistle-blowing, it should be a last resort, occurring only once certain conditions are met.

REVIEW QUESTIONS

1. Describe three characteristics/conditions of loyalty. Additionally, explain duties of loyalties engineers have and how these can come into conflict.
2. Define "legitimate" authority and describe three situations that would illegitimize the authority of an employer. Why is this concept important for engineers?
3. Describe the difference between actual, latent, and potential conflicts of interest and give examples of each.
4. Define whistle-blowing, explaining its relationship to loyalty in the process.
5. List and describe three conditions outlined by Richard DeGeorge that, if fulfilled, make whistle-blowing morally permissible and the additional two conditions that make it morally required.
6. Describe forms of dissent in corporate environments and the relationship between autonomy, dissent, and whistle-blowing and why these notions/actions are relevant to employed engineers.
7. Are engineers different from other professionals in their relation to their employers? If so, what are some of these differences?
8. Describe the nature of internal versus external whistle-blowing, giving examples of both.

REFERENCES

Bok, S. (1980). Whistleblowing and professional responsibilities. *Ethics teaching in higher education* (pp. 277–295).

Callahan, E., & Dworkin, T. (1992). Do good and get rich: Financial incentives for whistleblowing and the false claims act. *Villanova Law Review, 37*, 273.

Castellina, V. (2011). New financial incentives and expanded anti-retaliation protections for whistleblowers created by Section 922 of the Dodd-Frank act: Actual progress or justice politics. *Brooklyn Journal of Corporate, Financial, and Commercial Law, 6*, 187.

Dandekar, N. (1991). Can whistleblowing be FULLY legitimated? A theoretical discussion. *Business and Professional Ethics Journal, 10*(1), 89–108.

Davis, M. (1983). Some paradoxes of whistleblowing. *Business and Professional Ethics Journal, 2*(2), 39–47. Winter.

Davis, M. (1989). Avoiding the tragedy of whistleblowing. *Business and Professional Ethics Journal, 8*(4), 3–19.

Davis, M. (1998). Conflict of interest. *Wiley Encyclopedia of Management, 2*, 1–3.

DeGeorge, R. (1999). *Business ethics* (5th ed.). Upper Saddle River, NJ: Prentice Hall.

Fleischer, H., & Schmolke, K. (2012). Financial incentives for whistleblowers in European capital markets law? Legal policy considerations on the reform of the market abuse regime. *Legal Policy Considerations on the Reform of the Market Abuse Regime; ECGI-Law Working Paper.* August.

Gunsalus, C. (1998). Preventing the need for whistleblowing: Practical advice for university administrators. *Science and Engineering Ethics, 4*(1), 75–94.

Huttl, T., & Lederer, S. (2013). Whistleblowing in central Europe. *Public Integrity, 15*(3), 283–306.

Lubalen, J., & Matheson, J. (1999). The fallout: What happens to whistleblowers and those accused but exonerated of scientific misconduct? *Science and Engineering Ethics, 5*(2), 229–250.

Luegenbiehl, H. (2005). Whistleblowing. In C. Mitcham (Ed.), *Encyclopedia of science, technology, and ethics*. Detroit, MI: Macmillan Reference.

Martin, M. (1992). Whistleblowing, professionalism, personal life, and shared responsibility for safety in engineering. *Business and Professional Ethics Journal, 11*(2), 21–37. Summer.

Martin, M., & Schinzinger, R. (2010). *Introduction to engineering ethics* (2nd ed.). New York, NY: McGraw Hill.

Mayer, R. (1977). I'm not that hungry—Yet! *IEEE Engineering Management Review, 1*(5), 66–68.

Vandekerckhove, W., & Commers, M. (2004). Whistle blowing and rational loyalty. *Journal of Business Ethics, 53*(1-2), 225–233.

Chapter 10

Issues of Broader Concern for Engineers

Chapter Objectives

Having read this chapter, completed the included exercises, and answered the associated questions, readers should be able to

- with regard to the case of Qihoo 360's P1 wireless router, explain why it is important for the public to have an adequate grasp of the ways technologies work and the responsibilities engineers have for this understanding;
- describe the reasons for and problems associated with the gulf in understanding that exists between the sciences-engineering and the humanities-public and what engineers might do to bridge this gap;
- outline the nature of and reasons for responsibilities that engineers, as a group, have to public participation, education, and engagement, giving examples from the case studies in this chapter;
- explain the relationship between laws and ethics, why strictly adhering to laws alone is insufficient, and how engineers might approach the law in international and cross-cultural contexts;
- discuss the importance of being sensitive to local needs in the formulation and implementation of international aid work related to engineering and technologies.

CASE STUDY ONE—ENGINEERING AND PUBLIC KNOWLEDGE: THE "PREGNANCY" MODE ON QIHOO 360'S P1 WIRELESS ROUTER

A Chinese man was reported as having gone door to door, asking his neighbors to turn off their wireless routers: his wife was pregnant, and he feared the wireless signals would harm his unborn child (Yu, 2015). Similarly, villagers from Dongguan, Guangdong province, China, staged protests forcing a mobile telecom company to shut down its base station: they also feared that mobile signals would adversely affect their health (Wu, 2015). These reactions could be understood as relating, in part, to the "fear marketing" of a wireless router. This case touches on issues of broader concern for engineers.

Global Engineering Ethics. http://dx.doi.org/10.1016/B978-0-12-811218-2.00010-2

In Jun. 2015, the Chinese technology company Qihoo 360 released the P1 wireless router. Qihoo advertised the router as having a "pregnancy" mode, a setting that would ensure the health of expecting mothers and their unborn children. To achieve this effect, the setting simply decreased the router's power level. Shortly after its release, Qihoo held a press conference and changed the name from "pregnancy" to "energy-saving" mode. These were responses to an onslaught of criticisms by both its competitors and the public: Qihoo executives had used the "pregnancy mode as a marketing tactic," intentionally playing on public ignorance to pursue profits (Kleinman, 2015).

Zhou Hongyi, the chairman and CEO of Qihoo, had established a hardware division within the company to compete with routers produced by Xiaomi, another major Chinese technology company. The development of Qihoo's router was announced on Nov. 25, 2013 (Zhang, 2013a). The decision to include a pregnancy mode was decided shortly thereafter, at the end of 2013. Although wireless and cellular devices emit radio-frequency electromagnetic fields, when designed and produced according to industry and legal guidelines, there is no evidence of health risks (Schofield, 2012; Shi, 2013).

Hence, Qihoo's marketing campaign for its router would be like one by Coca-Cola for "fat-free water." Just as water is already fat-free, so too are wireless routers harmless to pregnant women and fetuses. However, unlike water, the public is largely ignorant of the nature and consequences of new technologies. In addition to Zhou, a number of other individuals who oversaw the development of the P1 router would have known this.

Jiang Xuxian was responsible for wireless safety within Qihoo 360, and his expertise was in the field of wireless communication. Shen Haiyan was responsible for the new hardware division within the company, and Huang Xifeng was responsible for marketing at Qihoo 360. Although Huang oversaw advertising for the router, both Jiang and Shen acquiesced to the ploy, despite possessing the requisite professional knowledge to dispel such claims.[108] Insofar as Qihoo infringed on the rights of individuals "to obtain true information" regarding "the commodities they purchase and use," their action could be considered illegal ("Law," 1993). Even if the company did not act illegally, however, their actions could still be considered unethical.[109]

As has been discussed previously, in the contemporary world, engineers have great power and great responsibilities that come with this power. Although these include not engaging in illegal activities, this is far from a comprehensive account of the duties of engineers. In addition to the fact that laws can conflict with ethics, not even acting completely in accord with good laws fully encompasses the sphere of ethics. This chapter more fully considers the responsibilities engineers as a group have to the public, the relation between ethics and the law, and how engineers can engage with laws in international and cross-cultural contexts.

---

108. Regarding who was responsible for what, see Zhang (2013b).
109. Information in this case study is additionally based on Feng (2015) and Custer (2015).

EXERCISE ONE—ENGINEERING AND PUBLIC KNOWLEDGE (PART ONE)

With reference to the case of Qihoo 360's P1 wireless router, complete steps 1–10 of the case-study procedure, using all relevant principles reviewed thus far.

10.1 ENGINEERING AND SOCIETY: POSITIVE CONTRIBUTIONS

In many ways, the analyses up to this point have focused on restrictions to the activities of engineers—the limitations of their actions by engineering and business ethical principles. The present chapter considers broader activities in which engineers should be encouraged to engage, based on their roles in society.

Again, the first basic ethical principle for global engineering is as follows: "Engineers should endeavor, based on their expertise, to keep members of the public safe from serious negative consequences resulting from their development and implementation of technology." Thus far, this principle has been interpreted and applied in a relatively narrow fashion, used to refer to the responsibilities that arise from the connection that individual engineers have to the development of technologies. Perhaps an understanding of this principle in these terms arises naturally, as a result of conceiving engineering activities in relation to individual companies. However, technological development can be conceived in much broader terms. Thus, the focus here is on the involvement of engineers in wider contexts of public discussions of technologies.[110]

As citizens, all adults have responsibilities to participate in public life. Engineers, it could be argued, have greater responsibilities than those of ordinary citizens. This argument would be based on the expertise of engineers, again, where this expertise brings with it responsibilities. Engineers are experts in more than just the projects on which they happen to be working. Given their roles as employees, however, they tend to view themselves as simply servants following directions. Perhaps because of this, engineers are sometimes unwilling to assert their more general beliefs in public forums; this would be compounded by feelings of loyalty to their employers.[111]

These circumstances contribute to the fact that most public discussions concerning technologies are initiated and conducted by persons lacking the appropriate knowledge base, in other words, nonengineers. A situation such as this increases the possibility that poor decisions will be made or that engineers will ignore decisions made by nonengineers. Due to a disconnect between engineers

---

110. Concerning a shift in focus of engineering ethics—from "microethics," aimed at fostering the capacities of individual engineers to engage in ethical reflection, decision-making, and action, to "macro-" and "mesoethics," expanding to encompass the responsibilities of professional and social organizations, and focus on value systems—see Herkert (2001), Herkert (2005), and Bocong (2012).
111. For a fuller discussion of loyalty, see Baron (1984).

and the public, the potential for negative consequences increases greatly. A variety of perspectives compellingly demonstrate this claim[112]:

• Do you believe that, as citizens, all adults have responsibilities to participate in public life? Why or why not? Explain what public life means to you.

10.1.1   The Two Cultures View

In his famous *The Two Cultures*, Charles Snow discusses two fundamentally different approaches to conceiving the world, one by scientists and engineers and one by humanists (Snow, 2001). In large part, he wrote this book to help overcome divisions between the two cultures. Unfortunately, however, problems associated with these divisions seem to have become more acute since the publication of this work in 1959.

According to Snow, on the one hand, scientists conceive the world in rational terms, where the notion of progress and efficiency dominates. Scientists believe the future lies in their hands and that this future is filled with the wonders of technologies—a view generally associated with "technological optimism." On the other hand, the humanities conceive the world in intuitive terms, where reason can only provide partial and biased answers. The central concerns of the humanities are human meaning and the fulfillment of broader missions.

Each culture is suspicious of the other, believing justifications provided are simply means of attaining greater control. As a result of this mutual suspicion and the fact that these two cultures use fundamentally different vocabularies, they have had—and continue to have—significant difficulties communicating with each other. Communication is necessary, however, since the future involves everyone. This can only occur if each side is empathetic to the other, explicitly recognizing the value of the other.

10.1.2   The Traditional Versus Superculture View

Kenneth Boulding has stated this same problem in another way: in the modern world, two cultures exist in tension, traditional culture and superculture.[113] According to Boulding, on the one hand, traditional culture is composed of folk technology, an emphasis on nations, the dominance of religious values, and loyalties to ethnic identities. On the other hand, superculture is composed of technological development, English as a common language, formal education, equality, and scientific ideals. To ensure true

---

112. The following two accounts are by no means comprehensive pictures of conceptualizations regarding the relations between the sciences/engineering and the humanities/public. For other interesting accounts that touch on such issues, the reader is referred to the work of Bruno Latour, Isabelle Strengers, and Donna Haraway, for instance, Latour (2005), Stengers (2010, 2011), and Haraway (1991).
113. See especially Boulding (1969).

progress, claims Boulding, these two cultures must work together: whereas traditional culture cannot ensure progress, superculture alone does not possess an adequate value base to achieve its objectives. A balance is thus required between the two, where the transmission of traditional values is balanced with an emphasis on creativity.

10.2 THE RELATION OF ENGINEERS TO THE PUBLIC: PRINCIPLES OF INVOLVEMENT

To achieve both material progress and human fulfillment, adequate interaction between scientific/engineering and humanistic concerns is essential—greater understanding of each by the other. Assuming the public has considerable difficulties understanding the work of engineers, a great deal of this responsibility for communicating and interacting falls to engineers. In recognition of this fact, several principles follow.

Unlike the previous principles, however, these would not be ethically required of individual engineers. Strictly speaking, individual engineers are only ethically accountable for their own work.[114] As a group, however, engineers have broader responsibilities to the public through their employment of technology. "Technology" is an ambiguous terms, referring to more than simply isolated sets of devices, processes, and techniques. The employment of the combinations of technologies—understood in broader terms—has significant systemic implications. Someone should assume responsibility. Again, for nonengineers, however, much of what occurs in relation to technologies is fundamentally obscure, and the lack of understanding precludes autonomous choice. As a whole, engineers should be involved in this process.

Given their role in society, as a group, engineers have a responsibility to establish communication with nonengineers, insuring that broader nontechnical concerns are considered. Simply because a group is responsible does not mean each of the individuals belonging to that group is equally responsible. Individual engineers cannot be ethically required to fulfill responsibilities associated with public participation, education, and engagement, resulting from a more general problem regarding the nature of collective guilt.[115] As a group, however, engineering should encourage individual engineers to participate in processes of communication and participation with the public, such that the responsibility of the group would be fulfilled. On this basis, three principles can be formulated.

---

114. For a fuller account of the nature of responsibility within engineering, see, for instance, Van De Poel, Fahlquist, Doorn, Zwart, and Royakkers (2012).
115. A situation where the group as a whole would be responsible, but not necessarily any one individual within that group, is known as one of "many hands." For a fuller discussion of this problem within the sphere of engineering, see Van De Poel et al. (2012) and Van de Poel, Royakkers, and Zwart (2015). Within the political domain, see Thompson (2004).

10.2.1   Principle of Public Participation: Engineers Should Seriously Consider Participating in Public Policy Discussions Regarding Future Applications of Technology

Again, the basis of this principle is the expertise of engineers, bringing with it responsibility. Engineers should be encouraged to actively engage in public discourse surrounding technologies, as opposed to doing so only when requested—since that might never occur or occur only infrequently. Where their input would be appropriate, engineers should seek out opportunities to engage in public discussions. However, participation should be understood appropriately.

"Participation" does not mean that engineers should make final decisions about the implementation of technologies independently of other stakeholders. A process such as this would result in "technocracy," government by those responsible for technologies, based on the technological ideals of rationality and efficiency. Since decisions about the future of technologies involve more than simply technological factors, this would be an unfortunate outcome.

As discussed before, the most important of these factors is the role of social values, about which engineers generally know no more than other members of society. Additionally, in relation to technological fields, other members of society, such as scientists, have expertise equivalent to that of engineers. A consensual approach to decision-making is, therefore, the most appropriate, where the interests of all affected stakeholders are considered.

10.2.2   Principle of Public Education: Engineers Should Seriously Consider Helping the Public to Understand the Applications of Technologies in Broader Social, Global Contexts

Again, technologies never exist in isolation: they exist in specific cultural contexts and tend to have global effects. A contextual understanding of technologies is necessary to adequately understand their implications and effects. At present, generally, neither engineers nor members of the public have such understandings. Engineers typically view technologies in terms of their technological contexts and immediate effects. Members of the public generally have only a vague understanding of technological processes. Education is thus necessary to overcome this divide. The question then arises as to where such education should be directed.

Although not always the case, members of the public typically show disdain toward gaining expertise in technological understanding.[116] Evidence of this

---

116. This is by no means always the case. In the fields of science and engineering studies, works by the likes of Stephen Hawking, Carl Sagan, Henry Petroski, and Don Norman have been relatively successful in captivating general audiences. Additionally, despite whatever shortcomings the genre might have, TED talks (http://www.ted.com) provide a forum in which scientists and engineers can more easily communicate with the general public.

would be failures to captivate the interests of humanists through the establishment of science, technology, and society education programs. For this reason, the more fruitful option seems to be educating engineers to better understand the broader implications of technologies—engineering education involving the concerns of social sciences and humanities related to technologies.[117]

Given their expertise, engineers would be able to increase public understanding, since they should not be the only or ultimate decision-makers. Only an informed public can make appropriate decisions regarding technologies, and engineers could be a source of information to educate the public. Thus, a part of engineering education should show engineers the importance of this role, since it cannot be ethically required of individual engineers.

10.2.3   Principle of Engineering Engagement: Engineers Should Seriously Consider Becoming Involved in Helping to Improve the Technological Futures of Those Less Fortunate Than Themselves, on a Voluntary Basis

Individuals are inadequately served by technologies. The reasons for this include economic poverty, social rejection, and geographic remoteness. Although technologies are not the solution to all—or even many—social problems, their applications can undoubtedly improve human life. The need exists to fulfill the basic conditions of human life, and, as long as that need exists, a corresponding responsibility to fulfill that need.

The fulfillment of this responsibility—and alleviation of human suffering more broadly—requires a combination of various efforts, primarily those relating to finance and policy. Additionally, efforts should include the creation of technologies appropriate to cultural contexts. Here engineers are in the best possible positions to meet these efforts and, therefore, have a responsibility to do so.

As a group, engineers should encourage individual engineers to assist in the alleviation of social problems through the application of technologies. It should be kept in mind, however, that not all problems can be corrected with a "technological fix," and engineers are not the sole decision-makers.[118] To reiterate, engineers should be educated to develop technologies appropriate to given social contexts, not simply apply cutting-edge technologies that fail to meet human needs—a mistake that has been repeated time and again. Finally, it should be noted that this principle encourages volunteerism. This results from the fact that waiting on adequate financial support can often be a fruitless endeavor.

---

117. To a large extent, ABET's criteria for student outcomes aim at this end, especially outcome h, requiring that programs impart to students "the broad education necessary to understand the impact of engineering solutions in a global, economic, environmental, and societal context" (Criteria).
118. For a fuller discussion of what might and might not be expected of technologies, see Weinberg (2003).

Just as in the process of applying ethical principles in general, it is important to keep in mind that the implementation of the above principles requires the education of engineers in matters that are not strictly technical in nature, although these matters are related to technical concerns[119]:

- Although they have been explained as applying to engineers as a group, do you believe there would be instances in which the principles of involvement would be ethically required of individual engineers? Why or why not? Give examples to clarify/justify your answer.
- Explain instances in which you have used a "consensual approach to decision-making" in the course of your engineering education and/or work. Why is this type of approach especially important to the development of technologies?
- Although volunteerism is strongly encouraged for engineers, could paid projects produce similar results? Why or why not? Provide any relevant examples with which you are familiar.

10.3 ETHICS AND THE LAW: THEIR SIMILARITIES AND DIFFERENCES

In studying ethics, students are often puzzled by the relationship between ethics and the law, probably because many so closely identify the one with the other— assuming that by strictly following the law, one is also acting ethically. However, especially in international contexts, this would be a mistaken assumption.

On the one hand, laws vary greatly from place to place and time to time. Even in relatively homogenous cultures, different localities adopt different laws. On the other hand, laws and ethics have similar functions. Both function to help people get along with each other and establish behavioral expectations. However, the two differ in the range of actions they cover and their normative force. To begin to understand these differences, considering the rationales for the law is important.

In *Leviathan*, a famous work on "social contract theory," the English political philosopher Thomas Hobbes proposed the hypothetical "state of nature" (Hobbes, 1982). This state would lack rules to govern behaviors; everyone would be primarily concerned with his or her own interests. If no rules exist, then all actions furthering self-interests are permissible. For example, in the state of nature, if I wanted to take food from you by force because I was hungry, then I would be justified in doing so. Obviously, however, you would also be justified in doing the same to me. As a result of this absence of rules, claimed Hobbes, human life would be "short, nasty, and brutish," since respect for the lives of others in the pursuit of self-interests would not be a requirement. Hobbes believed that all would ultimately recognize that this mode of existence is not

---

119. For more on the broader, positive duties professions have to society, see Davis (2002).

beneficial, appointing a strong authority—a sovereign, the Leviathan, or something like a dictator—who would provide control and, therefore, protection.

Even without the appointment of a dictator, however, it is obvious that human interactions should be governed in a manner to avoid chaos. Toward this end, violations of laws are associated with forms of punishment. Laws, thereby, establish minimum standards all are expected to follow.[120] Ethics, by contrast, involves more than simply minimum standards. It involves care and concern with the effects of actions on the lives of others. Having different foundations, laws and ethics are instantiated in different ways. The following is a brief list of the ways laws and ethics can differ, some of which have been discussed previously:

- Laws establish minimum standard for behaviors and are legally enforceable. By contrast, ethics establishes standards for good—rather than simply minimally acceptable—behaviors and might not be legally enforceable.
- Laws are instruments of compromise arrived at through legislative decision-making, while ethics aims at the good and is based on the use of reason.
- Laws can be immoral and amoral, since anything can be made legal or illegal in a given social context, while ethics are concerned with actions that have the potential to seriously affect the lives of others, thereby applying to a potentially more limited domain.
- Laws are slow to develop, covering new sets of circumstances, while individuals can reflect on and reason about ethics at any time.
- Laws vary from society to society, while ethical principles should provide relatively consistent results, or differences in their results should be principled.

- For each of the instances listed above, give examples of ways laws and ethics differ. Where necessary, be sure to clarify/explain your examples.

10.4 ENGINEERS AND LAWS, INTERNATIONALLY AND CROSS-CULTURALLY

Having discussed the nature of laws, it is necessary to consider the relationship engineers should have to the law, especially in international and cross-cultural contexts, since laws vary and might be immoral. Given these conditions—and the previously outlined duties of engineers and employees—engineers might approach the law as follows:

1. Engineers have a *prima facie* (as an initial response) duty to obey laws. If no countervailing considerations exist, then this becomes an actual duty, regardless of the country or locality in which engineers are working.
2. Engineers should obey all local laws, assuming these laws do not have moral implications, in other words, that obedience to such laws would not have the

---

120. This is a conception of law based on "retributive justice."

potential to seriously harm others. Laws that strictly embody local customs and traditions, for example, should be obeyed.

3. When engineers are involved in several local contexts simultaneously, they should obey the strictest version of relevant nonmoral laws. In other words, if one set of laws captures all the major features of a second one but the second one does not capture all the major features of the first, then the first set of laws should be followed. In circumstances where nonmoral laws conflict, however, some prioritization might be necessary. In situations such as these, engineers should consider the specific set of circumstances in which they find themselves or adopt an additional principle, such as following the laws of the locality in which they are physically present.

4. Engineers should do their best to refuse to follow local laws that violate business and engineering ethical principles. The phrase "should do their best to refuse" is included to account for mitigating circumstances—an example of which is mentioned below. If no such conditions exist, however, then engineers would not be justified in violating ethical principles. In many cases, engineers will have to make individual judgments regarding the extent to which circumstances could be considered conditions excusing one from adhering to ethical principles.

5. Forms of coercion by local governments constitute mitigating circumstances, conditions that would excuse engineers from adhering to ethical principles.

6. Engineers should use their influence in local contexts to change laws that violate ethics. Even in circumstances where engineers have excusing conditions for violating ethical principles, they have obligations to try to bring about change.

- How might engineers use their influence in local contexts to change laws that violate ethics? List any examples of this with which you are familiar.

EXERCISE TWO—ENGINEERING AND PUBLIC KNOWLEDGE (PART TWO)

Referring back to the work you did in exercise one on Qihoo 360's P1 wireless router, reconsider steps 6–10 of the case-study procedure, noting any additional principles or concepts that might apply to your analysis of this case.

CASE STUDY TWO—ALIGNING INTERNATIONAL RESPONSIBILITY WITH LOCAL NEEDS: ENGINEERING AID WORK

Service in Engineering: Opportunities and Challenges

This chapter has highlighted some of the positive responsibilities of engineers, one of which is employing technologies to improve the situations of those in need. Engineering associations and nongovernmental organizations (NGOs) engage in such activities collectively, organizing and pooling resources. These engagements can face particular difficulties.

In engineering aid work, projects are often initiated and planned by people living in places and circumstances different from the anticipated beneficiaries of these projects, resulting in a potential disconnect between those designing technologies and those using them. Additionally, aid projects are generally funded with other people's money—donations or government funds collected through taxation. Efforts should thus be made to ensure that resources are allocated in an efficient manner—to maximize the benefits and minimize the costs associated with such projects. Insofar as engineers possess professional expertise, from a technological perspective, they are in a position to assess the anticipated benefits and costs of these projects.

To better understand the nature of engineering aid work and ethical issues associated with this work, the following case study examines three projects. Since water, energy, and food provide the very conditions of life—and the pursuit of these resources has often affected the environment in negative ways—the focus here is on projects related to the acquisition of these resources in Malawi, Nepal, and Indonesia. In addition to particular instances of good and bad, these projects also draw attention to more general problems associated with aid work and potential solutions. These include the potentially divergent objectives of those who initiate and provide funds to engineering aid projects and the beneficiaries of such projects and the importance of sensitivity to local practices and needs.

Water in Malawi: From Failure to Long-Term Sustainability

In 2009, Engineers without Borders, Canada, an engineering-related NGO, completed a gravity-fed water system in Machinga, Malawi. The systems consisted in a series of pipes running from clean water higher up in the mountains to water taps in the village. Just over a year after the completion of this project, less than half of the water taps installed still functioned. Although disheartening, this situation is not unique: many similar projects have been started—and failed—in this same rural village.[121]

Despite its proximity to Lake Chiuta, Machinga lacks a reliable source of clean water. Women and children often walk several hours to collect drinkable water. A variety of NGOs and nonprofit organizations worked to develop and implement solutions to Malawi's water problems and, in 2009, Engineers without Borders Canada (henceforth EWBC[122]) took on the task.

---

121. Perhaps the situation is not as bleak as it might initially seem: regarding the centrality of failure to innovation within engineering, see Petroski (1985).

122. Engineers without Boarders, Canada, falls under Engineers without Borders (hereafter EWB), an umbrella organization of NGOs. For more information on EWB, see their website at http://www.ewb-usa.org.

After the completion of the project, EWBC sent a team to evaluate the system, discovering many of the taps were not working, 81 of 113. Community members had attempted to patch pipes and fix broken taps, but due to a lack of available parts, they were unable to make the repairs. In addition to broken taps from the EWBC project, the team discovered taps from similar projects sponsored by other NGOs that had failed shortly after their completion.

Whether the installation of water taps or solar energy panels, consideration of long-term sustainability is one of the major problems with such projects. EWBC did not plan for either the long-term maintenance or the funding of the system. The funding structures of such projects present unique challenges as well. Unlike projects in either the private or corporate sector, the donors for aid projects do not necessarily have direct ties to the beneficiaries of these projects. As a result, the projects are often designed according to short-term interests of the donors rather than the long-term interests of the beneficiaries. This is common within the developmental sector, raising issues regarding the design and implementation of such projects.

Such questions have prompted changes in the ways such organizations deal with failures: EWB now publishes an annual failure report, so others can learn from and avoid similar mistakes, essential to the long-term sustainability of engineering aid work.[123] In recent years, an emphasis on user-centered design is another strategy employed to avoid or minimize failures associated with engineering aid projects:

- In your opinion, who is primarily responsible for the failure of this project? Justify your response.
- Should engineers have responsibilities to work on projects that benefit people geographically so far removed from themselves? Why or why not?

Biogas Digesters in Nepal: The Importance of Culture in User-Centered Design (Successes)

In 2008, the student chapter of EWB at the Technion—Israel Institute of Technology (hereafter Technion) worked closely with the 8000-person village of Namsaling, Nepal, to improve the design and creation of biogas digesters—devices used to produce gas for cooking and fertilizer for growing. Through consultation and work with local stakeholders, the student team gained a sense of the culture, needs, and values of those in Namsaling, resulting in the implementation of less costly digesters.

Biogas digesters can solve a variety of energy, environmental, and health issues in rural communities. Biowaste—animal and sometimes human excrement, along with weeds or other biological waste—are placed in the digester.

---

123. Based on Damberger (2011), Kampala, Chigwenembe, and Rabbani (2009), Mpaka (2015), Mpumulo (2014), Schmidt (2010), and Scott (2012).

As the material decomposes, it produces gas that can be pumped out and used as cooking fuel and to heat homes. The remaining waste can be used as a fertilizer for local farms and gardens. This gas reduces the amount of wood burned, therefore reducing deforestation and exposer to the toxic smoke and fumes produced by traditional wood-burning stoves, in turn reducing respiratory problems (Tugend, 2011). Therefore, the use of natural gas from biogas digesters contributes to environmental sustainability and the health of rural communities.[124]

When the Technion student project began in 2008, Nepal already had 200,000 biogas digesters in use (Stricker, 2010), although their construction cost considerable time and money. To build a digester of this kind, a large pit was dug and soil used to create a dome, often by the children of the community. Concrete was then laid over the dome, and soil removed to create the main chamber of the digester. Digging the pit, fashioning the soil dome, laying the concrete, and removing the soil afterwards required tremendous time and effort. Technion students believed they could design a cheaper, easier way of building biogas digesters.

Students made several trips to the village, "to work with villagers in defining and collecting data required for the design of sustainable and appropriate projects" (Lichtman, n.d.). Additionally, they worked closely with the Namsaling Community Development Center (NCDC), Biogas Sector Partnership (BSP), and families from Namsaling. The NCDC and families in Namsaling not only assisted in the development of a solution but also funded a third of the project (Tugend, 2011). With these partners, the team redesigned the dome mold used in the construction process.

Instead of using soil to build the dome, the team used bamboo—a material widely available in Namsaling. After the concrete on top hardened, the mold could be removed from the pit more easily and reused to create other digester chambers. This significantly reduced the amount of money and time needed to create digesters. After this new design was implemented, villagers reported a 36 kg reduction in daily wood use per family (Lichtman, n.d.). The Technion team's "deep acquaintance with the community" in Namsaling undoubtedly contributed to the success of this project (EWB, n.d.).

Students from Technion spent considerable time with the community stakeholders, learning more about their values and economic needs, for example, the Nepalese emphasis on family and respect for community members, which was important to the construction and use of the digesters. Additionally, the economic benefits associated with digester fertilizer are particularly significant: the fertilizer is used in Namsaling to grow important cash crops such cardamom and ginger (Namsaling village, n.d.). The success of this project emphasizes the importance of considering the values, needs, and circumstances of the Nepalese

---

124. See Forte (2011) for additional information on the use of biogas digesters in Nepal.

people affected by engineering aid projects. Even with the best of intentions, neglecting these considerations can result in the failure of engineering aid projects:

- In your opinion, who is primarily responsible for the success of this project? Justify your response.
- Did the financial contributions of the Namsaling community help this project to succeed? Why might the financial support of beneficiaries be helpful to the success of a project—versus projects funded entirely by external organizations?

Cook Stoves in Indonesia: The Importance of Culture in User-Centered Design (Failures)

To prepare food, make medicine, and heat water and homes in the winter, households in rural communities rely on open-fire cook stoves. However, these stoves cause various problems to the environment and to the health and safety of those using them. In Indonesia alone, an estimated 24.5 million households depend on these stoves, resulting in 165,000 premature deaths each year from stove-related air pollution. Additionally, women and children spend hours gathering wood to burn each day, taking up time and contributing to deforestation.

Different NGOs have worked to develop safer, more energy-efficient stoves, but the adoption and therefore effects of new stoves have been limited. In the last 50 years, redesigned stoves have been distributed to almost 830 million people worldwide, but studies estimate the number of people relying on traditional, open-fire cook stoves will not change in the next 15 years. These findings highlight the importance of considering local needs and circumstances in the formulation of engineering aid projects, and considering the case of an Indonesian government initiative is instructive.

To reduce household reliance on kerosene and biomass, in 2007, the Indonesian government launched a program encouraging a switch to the use of liquefied petroleum gas (LPG) stoves, which pollute less and are more efficient than traditional stoves. Despite providing households with free start-up kits that included both a stove and LPG, a survey taken a few years later found that, although many used the LPG stoves, over 50% of households also used the traditional, woodburning stoves. In designing this program, the Indonesian government failed to consider all the needs of households and the ways they used the stoves.

Families generally use different stoves for different tasks, from cooking food to heating water. Some of these tasks are done inside, while others are done outside. Some of the new stoves were not mobile enough to be used both inside and outside. Additionally, larger families generally cook larger meals to feed more people, and the new stoves simply did not have the capacity to do so. This lack of attention to the social dimensions of technology contributed to the failures of the LPG stove project[125]:

---

125. Based on Aristani and Desa (2015), "Building Capacity," (2015), "Indonesia," (2013), Johnson and Bryden (2013), and "Clean Stove," (2014).

- In your opinion, who is primarily responsible for the failure of this project? Justify your response.
- Aside from those discussed above, list any engineering aid projects with which you are familiar. Would you consider these projects successes or failures? Why? What factors do you think contributed to the successes or failures of these projects?

Conclusion: The Benefits and Costs of Globalization in Engineering Aid Work

Technologies have the capacity to affect human life and the environment as never before, and engineers should strive to employ technologies to help those in need. As part of this trend, teams of engineer from one part of the world will increasingly work to design and implement engineering projects as solutions to the problems of groups in other parts of the world. While these trends in globalization have great benefits—improving processes and products by bringing more, different types of ideas and people together—they can also have costs.

Technical know-how alone in designing and implementing engineering aid projects is insufficient. Consideration should be given to the needs, values, and circumstances of the people engineering projects are meant to help. Geographic distances and cultural diversity are also features of globalization, and they can inhibit gaining knowledge of these needs, values, and circumstances. To move forward as a global society, however, acquiring such knowledge and understanding its impact on the adoption of technology is more necessary than ever.

EXERCISE THREE—ALIGNING INTERNATIONAL RESPONSIBILITY WITH LOCAL NEEDS

With regard to one of the cases above, complete steps 1–10 of the case-study procedure, using all relevant principles reviewed thus far.

10.5   SUMMARY

The case of Qihoo 360's P1 wireless router highlights problems associated with the public's ignorance of technologies. Although individual engineers have ethical duties associated with the work they perform, as a group, engineers also have responsibilities to the public—participating in public policy discussions on technologies, educating the public on the natures and consequences of technologies, and engaging with technologies to serve those in need. These responsibilities follow from the gulf that exists between understandings of technologies by engineers and the public, and the professional expertise of engineers. Although laws can aim to address and correct these problems, laws are not the same as ethics, and alone laws are incapable of addressing and correcting such problems. For these reasons, it is important that engineers be cognizant of differences between laws and ethics and how to engage with laws, especially in international and cross-cultural contexts. As the case of Aligning International Responsibility with Local Needs makes clear, in carrying out aid work, engineers should take into consideration the needs, values, and circumstances of the beneficiaries of these projects. This is especially true in international, globalized environments, where a failure to do so can result in wasted time, effort, and money.

REVIEW QUESTIONS

1. Describe three ways engineers abused their power and ignored their responsibilities toward the public in the case of Qihoo 360's Pl wireless router.
2. With reference to the work of Charles Snow, explain the difference between how the sciences conceive the world and how the humanities conceive the world.
3. Explain the differences between the traditional and superculture views, as outlined by Kenneth Boulding. Do you think this perspective is more accurate than that of Snow? Why or why not?
4. Describe one broader responsibility that engineers have to the public and why this responsibility is the basis for a principle of involvement.
5. Provide an example of engineers fulfilling each of the principles of involvement described above. Why should these apply to groups of engineers rather than individual engineers?
6. Explain three differences between laws and ethics and provide examples of each.
7. List and explain two circumstances contributing to the fact more engineers are not involved in public discussions concerning technology. What consequences could this lack of involvement have on the public?
8. What is a major difference between the principles outlined in this chapter and the principles outlined previously? Why is this significant?
9. How should engineers go about educating the public? Give two examples.
10. Why should engineers disregard local laws that violate engineering ethical principles? At what point does a law cease to be ethical? Again, give two examples.
11. With reference to the second case of Aligning International Responsibility with Local Needs, list two reasons for successes and two reasons for failures of engineering aid projects.

REFERENCES

Aristani, C. & Desa, Y. (2015). Session 3: Transition to cleaner cooking fuels and technologies: Impact on rural communities. *Sustainable Dissemination of Improved Cookstoves: Lessons from Southeast Asia. Workshop Report 13. Yangon, Myanmar*. December http://e4sv.org/wp-content/uploads/2016/03/WR13-Sustainable-Dissemination-of-Improved-Cookstoves-Lessons.pdf.

Baron, M. (1984). *The moral status of loyalty*. Center for the study of ethics in the professions Chicago: Kendall Hunt Publishing Company.

Bocong, L. (2012). From a micro-macro framework to a micro-meso-macro framework. In S. Christensen, C. Mitcham, B. Li, & Y. An (Eds.), *Engineering, development, and philosophy: American, Chinese, and European perspectives*. Dordrecht: Springer.

Boulding, K. (1969). The interplay of technology and values: The emerging superculture. In K. Baier & N. Rescher (Eds.), *Values and the future*. New York: Free Press.

Building Capacity and Sharing Lessons on Fuels in Indonesia. (2015). *Global Alliance for Clean Cookstoves*. 21 June. http://cleancookstoves.org/about/news/06-21-2015-building-capacity-and-sharing-lessons-on-fuels-in-indonesia.html.

Clean Stove Initiative Indonesia. (2014). StovePlus. http://www.stoveplus.org/en/projects/clean-stove-initiative-indonesia.

Criteria for Accrediting Engineering Programs, 2016–2017 (n.d.). *ABET*. http://www.abet.org/accreditation/accreditation-criteria/criteria-for-accrediting-engineering-programs-2016-2017/.

Custer, C. (2015). Chinese tech company exploits baseless superstitions with 'pregnancy safe' wifi router. *Tech in Asia*, 23 June https://www.techinasia.com/disgusting-chinese-tech-company-exploits-baseless-superstitions-pregnancy-safe-wifi-router/.

Damberger, D. (2011). *What happens when an NGO admits failure*. TED.com. April https://www.ted.com/talks/david_damberger_what_happens_when_an_ngo_admits_failure.

Davis, M. (2002). *Profession, code, and ethics*. Aldershot, Harts/Burlington, VT: Ashgate.

Engineers without Borders. (n.d.) http://www.ewb-usa.org.

EWB Technion—Nepal Group. (n.d.). *Engineers without Borders Israel*. http://www.ewb.org.il/#!ewb-technion-nepal/czqq.

Feng, C. (2015). China's Qihoo 360 launches wi-fi router to 'protect' pregnant women from radiation. *South China Morning Post*, 22 June http://www.scmp.com/tech/social-gadgets/article/1823880/chinas-qihoo-360-launches-wi-fi-router-protect-pregnant-women.

Forte, J. (2011). A glimpse into community and institutional biogas plants in Nepal. *Engineers without Borders*, . http://www.ewb-usa.org/files/2015/05/biogas-plants-in-nepal.pdf.

Haraway, D. (1991). *Simians, cyborgs, and women*. New York: Routledge.

Herkert, J. (2001). Future directions in engineering ethics research: Microethics, macroethics and the role of professional societies. *Science and Engineering Ethics*, *7*(3), 403–414.

Herkert, J. (2005). Ways of thinking about and teaching ethical problem solving: Microethics and macroethics in engineering. *Science and Engineering Ethics*, *11*(3), 373–385.

Hobbes, T. (1982). *Leviathan*. New York: Penguin Books.

Indonesia—Toward Universal Access to Clean Cooking. (2013). *The World Bank. Asia Sustainable and Alternative Energy Program*. June. https://cleancookstoves.org/binary-data/RESOURCE/file/000/000/361-1.pdf.

Johnson, N., & Bryden, K. (2013). Clearing the air over cookstoves. *Dem + Nd—ASME Global Development Review*. http://www.asmedemand.org/demand/fall_2013?pg=1#pg1.

Kampala, R., Chigwenembe, A., & Rabbani, E. (2009). *Water point monitoring system in Machinga district, Malawi*. Engineers without Borders Canada. September. http://blogs.ewb.ca/africanprograms/files/2010/08/WPMonitoring-in-Machinga-Final-Report-Sep-2009.pdf.

Kleinman, Z. (2015). 'Pregnancy wi-fi' router causes controversy in China. *BBC*, 23 June. http://www.bbc.com/news/technology-33237860.

Latour, B. (2005). *Reassembling the social: An introduction to actor-network theory*. New York: Oxford University Press.

Law of the People's Republic of China on Protection of Consumer Rights and Interests. (1993). *lawinfochina.com*. http://www.lawinfochina.com/display.aspx?lib=law&id=6137&CGid=.

Lichtman, G. (n.d.). Technion's engineers without borders. *Technion External Relations & Resource Development*. http://pard.technion.ac.il/technions-engineers-without-borders/.

Mpaka, C. (2015). Machinga residents lied to and used on water project. *The Times*, 21 September. http://www.times.mw/machinga-residents-lied-to-and-used-on-water-project/.

Mpumulo, A. (2014). Malawi: Water and sanitation poverty despite closeness to Lake Malawi. *All Africa*, 23 December. http://allafrica.com/stories/201412240147.html.

Namsaling Village Development Committee. (n.d.). Sustainable Development Periodic Plan Namsaling VDC. http://www.colorado.edu/mcedc/sites/default/files/attached-files/sdp_namsaling.pdf.

Petroski, H. 1985. *To engineer is human*. New York: St. Martin's Press.

Schmidt, G. (2010). Engineers without borders Canada 2009 Failure Report. *Engineers without Borders Canada*. https://www.ewb.ca/sites/default/files/2009 EWB Failure Report.pdf.

Schofield, J. (2012). Wi-Fi: Are there any health risks? *The Guardian*, 27 September. http://www.theguardian.com/technology/askjack/2012/sep/27/wi-fi-health-risks.

Scott, O. (2012). Near-term success, long-term failure. *Admitting Failure*, 12 January. https://www.admittingfailure.org/failure/owen-scott/.

Shi, P. (2013). Wifi won't hurt your health. [in Chinese] *People's Daily*, [in Chinese].

Snow, C. 2001. *The two cultures*. London: Cambridge University Press.

Stengers, I. 2010. *Cosmopolitics I*. R. Bononno (trans.) Minneapolis: University of Minnesota Press.

Stengers, I. 2011. *Cosmopolitics II*. R. Bononno (trans.) Minneapolis: University of Minnesota Press.

Stricker, L. (2010). Technion engineers lend a hand in Nepal. *The Canadian Jewish News*, 16 December. http://www.cjnews.com/news/technion-engineers-lend-hand-nepal.

Thompson, D. 2004. *Restoring responsibility: Ethics in government, business, and healthcare*. Cambridge: Cambridge University Press.

Tugend, T. (2011). Technion professor engineering a social conscience in developing countries. *Jewish Journal News*, 14 June. http://www.jewishjournal.com/community/article/technion_professor_engineering_a_social_conscience_in_developing_countries.

Van De Poel, I., Fahlquist, J., Doorn, N., Zwart, S., & Royakkers, L. (2012). The problem of many hands: Climate change as an example. *Science and Engineering Ethics*, *18*, 49–67.

Van de Poel, I., Royakkers, L., & Zwart, S. 2015. *Moral responsibility and the problem of many hands*. New York: Routledge.

Weinberg, A. (2003). Can technology replace social engineering? In E. Katz, A. Light, & W. Thompson (Eds.), *Controlling technology: Contemporary issues* (pp. 109–116). Buffalo, NY: Prometheus Books.

Wu, C. (2015). The communication base station is boycotted by villagers [in Chinese]. *Sina*, 15 July. http://tech.sina.com.cn/t/2015-07-22/doc-ifxfaswf8410684.shtml.

Yu, Z. (2015). A man asks every neighbor to shutdown their WiFi because his wife is pregnant [in Chinese]. *XinHua Press*, 15 April. http://news.xinhuanet.com/legal/2014-04/15/c_126389670.htm.

Zhang, N. (2013a). 360 start working on wireless area [in Chinese]. *Sina Technology*, 25 November. http://tech.sina.com.cn/i/2013-11-25/21418947140.shtml.

Zhang, N. (2013b). 360 adjusts the organization, Zhou Hongyi is responsible for the safety of wifi [in Chinese]. *Sina Technology*, 28 November. http://tech.sina.com.cn/i/2013-11-28/13548957646.shtml.

Chapter 11

The Rights of Engineers

Chapter Objectives

Having read this chapter, completed the included exercises, and answered the associated questions, readers should be able to

- explain the relationship between duties and rights and the nature of/basis for rights with regard to moral, civic, and employment communities;
- describe the nature and give examples of both racial and sexual discrimination, and the relation between rights and discrimination;
- list the rights of employees in general and engineers specifically, and explain problems with and solutions to the enforcement of rights.

EXERCISE—SEXUAL HARASSMENT IN THE WORK PLACE: JOAN MENDOZA

Read through each of the following hypothetical scenarios. Take a few minutes to respond to the issues posed after each section. If relevant, mention any ethical principles encountered thus far to justify your answers. If none of the principles reviewed thus far appear relevant to answering the questions, then briefly justify your answers, referring to principles you think would be relevant.

Joan Mendoza recently graduated from a state university with a bachelor's degree in civil engineering. She is quite pleased with herself, since Joan successfully completed a program without female professors and very few enrolled female students. Although she is quite attractive, Joan has developed many of the more "macho" attitudes of her fellow civil engineering students, in part as a result of the environment of her school. Joan is confident a successful career as a civil engineer awaits her, in part because she had a number of promising job offers well before graduation. Since Joan wants a stable work environment, and the opportunity to give back to her community and work outdoors, she accepted a position as a project engineer with the State Highway Department (SHD).

Her interviewer painted a glowing picture of her career prospects with SHD. After her interview, Joan also spoke with another female engineer who was enthusiastic about her work. During the interview, no one mentioned to Joan that she would be only the second female engineer hired by SHD and the first

Global Engineering Ethics. http://dx.doi.org/10.1016/B978-0-12-811218-2.00011-4

female project engineer. Joan had not thought to raise any questions concerning this subject:

1. Should the interviewer have explained to Joan her pioneering role at SHD? Why or why not?
2. Should she have raised the question of the number of female engineers working for SHD?
3. Do you think the failure of the interviewer to mention the number of female engineers working for SHD was planned? Explain your answer.

Joan thoroughly enjoys her first several months at work, which consist in introductions to various parts of SHD and more generalized training. She is suddenly sent to a rural part of the state to oversee a road repair project, after the previous manager developed a serious long-term illness. Joan packs her bags with the anticipation of facing new challenges. When she arrives at the project site, her subordinates behave in a rather hostile fashion, which Joan dismisses as a sign they are unprepared for the sudden change of leadership. However, she quickly discovers she is the only woman onsite. After several days, workers begin making comments: "Bitches belong in the house, not a construction site!" "She's not worth much as an engineer, but I'd like to get her alone..."

1. Do you think these comments are normal and to be expected at construction sites or a normal reaction to changes in leadership? Explain your answer.
2. Are these comments demeaning to Joan as an engineer or as a woman?
3. Should she confront those making the comments? If so, then should Joan order them to cease making such comments and/or report them to the headquarters of SHD? Should she request a transfer?

Despite the comments, Joan is able to focus on her job, and after a couple weeks, the comments cease. In their place, however, she notices that—whenever she leaves her office—the men stare at her intently, almost longingly. One afternoon, just before work ends, one of the men, Bill Smith, hesitantly approaches Joan. Stuttering, Bill says he is attracted to her and would like to take her out for a drink. Joan finds his demeanor generally appealing and is tempted to take him up on the offer:

1. How should Joan react to Bill's invitation? Explain your response.
2. What limits exist within professional relationships?

Later that month, Joan notices something else: whenever she is consulting with the foreman, Tom Bryant, he seems to get very close to her. Sometimes his body touches hers. At first, Joan assumes this is an accident, but as time passes, the touching becomes more frequent. She thinks he is intentionally rubbing against her:

1. How should Joan react to Tom's behavior? Justify your answer.
2. Does Tom's behavior constitute sexual harassment? Why or why not?

Several months later, Joan's supervisor, Bob Hamm, arrives at the site to inspect the project. After the review, during a private consultation in her office, Bob tells Joan the project seems to be falling behind schedule, and he will have to write a negative performance review of her work. As a married man, Bob says he understands that women sometimes have difficulties adjusting to new environments and that—were they to spend the night together—he could help her adjust. Bob tells Joan that if she agrees, then he will revise his report on her work. As this is her first performance review, Joan knows it will be very important to her long-term career prospects with SHD:

1. Should Joan sleep with Bob? Why or why not?
2. There were no witnesses to this conversation. How important is this fact? Why?
3. Should Joan lodge a complaint with SHD's headquarters? Explain your answer.
4. If she does and SHD does nothing, then should Joan lodge a complaint with the federal Equal Employment Opportunity Commission? Why or why not?

11.1   NOT ONLY DUTIES, BUT ALSO RIGHTS

Up to this point, the focus has been largely on the responsibilities and duties of engineers, companies, employers, and employees. However, the flip side of such duties is rights. Western ethical and political thought has stressed rights to a greater extent than responsibilities; this is less true in Asian societies. Given this emphasis, especially in the United States, many have argued that too much stress has been laid on rights.[126]

Perhaps this emphasis can be explained with reference to the US Constitution, where the first 10 amendments are collectively known as the "Bill of Rights." These amendments have been understood as limiting the powers of the Federal government over individuals. For example, the well-publicized second amendment is generally understood as restricting the powers of the Federal government to limit weapon ownership. Again, especially in the United States, many citizens focus mainly on their rights and very little on their responsibilities and duties.

Given this state of affairs, the fact that relatively little emphasis has been given to the rights of engineers is interesting.[127] Perhaps this results from the fact that rights and obligations are correlative: if engineers have obligations in relation to one party, then rights are implied in relation to another. For example, as was discussed in Chapter 9, if engineers have important obligations to foster and preserve public safety, then they also have rights to violate employment

---

126. For examples of philosophers who have argued for a shift away from rights to more capabilities-based approaches characteristic of virtue ethics, see Nussbaum (2003) and MacIntyre (2013).
127. For notable exceptions, see Whitelaw (1975) and Flores (1980).

confidentiality to do so. Thus, it could appear as though discussions of rights are unnecessary, since they are already implied in discussions of duties. Since rights violations can be common, however, focus on them is necessary.

11.2 THE NATURE OF RIGHTS

A "right" can be understood as an entitlement of an individual. "Positive" rights are when one is entitled to others acting in a certain way, doing something. For example, a right to education would be a positive right, entailing that schools are built, teachers are hired, and so on. "Negative" rights are when one is entitled to others not acting in a certain way and not doing something. For example, a right to free speech would be a negative right, entailing that no one prevents one from speaking, writing, and so on. Obviously, positive and negative rights can be connected, for example, a positive right to medical treatment and a negative right to choose a physician to consult. Again, it is important to keep in mind that rights are correlated with duties and obligations: in the case of positive rights, the duty of one individual or group is to do something for another; in the case of negative rights, the duty of one individual or group is not to do something to another.

. Rights are central to the exercise of autonomy, since they guarantee the minimal conditions necessary to pursue interests. However, rights are not absolute. To protect the rights of others—or achieve an important social goal—rights can be restricted. These potential conflicts between the rights of individuals and society necessitate discussion regarding the ethical implications and consequences of rights.

Rights exist at various levels, the most fundamental of which are generally referred to as "human" rights. Since many have argued that rights at this level should be conceived as encompassing more than just human beings, calling them "moral" rights might be more appropriate.[128] In India, for example, dolphins have been declared "nonhuman persons," which prevents them from being kept in captivity for entertainment. Moral rights are based simply on belonging to a moral community. Membership in this community could be determined by possessing sentience, emotions, reason, and so on. It might also be determined by physical similarities to other beings.

With regard to dolphins, the Indian Ministry of Environment and Forest has stated that "various scientists who have researched dolphin behavior have suggested that the unusually high intelligence as compared with other animals means that dolphins should be seen as 'nonhuman persons' and as such should have their own specific rights, and it is morally unacceptable to keep them captive for entertainment purposes" ("India Bans," 2013). This does not mean, of course, that dolphins are entitled to the same rights and

---

128. Concerning claims that rights traditionally given to humans should extend further, see Regan (2004).

protections as human beings. Rather, because of a specific characteristic dolphins possess, intelligence, they are different from other animals and, therefore, entitled to rights—the right not to be kept for purposes of entertainment. Since one need only belong to the moral community to have moral rights, these have been understood as foundational, outweighing other forms of rights.[129]

Other forms of rights are contractual in nature, based on membership in particular groups: civic rights are based on membership in a particular nation or community. Employee rights are based on membership in a particular company, and the rights of engineers are based on membership in a particular profession and role responsibilities related to engineering. Insofar as these rights are contractual in nature, they are limited to particular relations among people. Civic rights, for example, do not transfer from one society to another, although moral rights do. For this reason, determining what counts as a moral right is especially important:

- Based on their level of intelligence, the Indian government has classified dolphins as "nonhuman persons," granting them certain rights. Should intelligence take precedence over other characteristics on the basis of which individuals and groups would be accorded rights? Why or why not?

The rights of engineers can be considered in terms of three levels: engineers as moral beings, as employees, and as engineers. As mentioned above, moral rights are the most fundamental, since they apply to all moral beings, in all situations. Again, moral rights override other types of rights. For example, employers do not have the right to violate the moral rights of employees to achieve business objectives, even if employees have signed contracts giving employers authority over them. These rights are crucially important, since they provide means of morally justifying actions and making demands. Moral rights can, therefore, be understood as trumps, to be invoked when fundamental aspects of people's humanity are threatened. Previously, it was explained that engineers have a duty to "ensure that fundamental rights will not be negatively impacted as a result of their work with technology."

Since persons have moral rights merely because they belong to the moral community, these rights reflect how persons deserve to be treated as members of that community. Therefore, in establishing moral rights, one must ask how moral beings deserve to be treated. This is clearly not a factual enterprise, since the treatment of persons varies socially and culturally. As a normative question, disagreement exists regarding which rights are truly moral. In part, this results from continuous attempts by stakeholders to equate civic with moral rights and to give the latter greater justification. Candidates for moral

---

129. Regarding the nature of human rights in general, see Ropp and Sikkink (1999) and Nickel (1987).

rights include—but are not limited to—a right to life, food, shelter, privacy, justice, nondiscrimination, free speech, fair treatment, and equal opportunities.[130] These are all based on the supposition that, in some sense, persons share a fundamental equality and should be able to act on their potentials as persons.

Of course, listing sets of rights is not the same as determining their meanings in concrete situations. For example, although a "right to life" refers to protection from the moment of conception for some, it refers to protection from the moment of birth for others. Additionally, although a "right to justice" could mean a trial before a jury composed of peers in some social contexts, it could mean the exercise of power for communal well-being by authority in other social contexts.[131] Hence, the meaning and status of different rights are much debated, and considerations and justifications of all these rights would go well beyond the scope of this text. Here it is important that engineers simply be aware of the nature of and debates concerning moral rights, for the sake of better understanding their specific professional duties and rights.[132]

11.3   EMPLOYEE RIGHTS AND THE LIMITS OF EMPLOYERS

As mentioned before, the rights of employees are based on either explicitly signed or implicitly understood contracts, agreed upon between employees and employers. These rights are, thus, specific to particular employment situations. For example, if an employee signs a contract to be paid a specific amount for the completion of a piece of work, then she has the right to be paid that amount once the piece of work has been completed. If an employee signs a contract to be paid at a rate of one and a half for overtime, then he has the right to be paid at a rate of one and a half for working overtime. Employee rights, thereby, limit the powers of employers to act arbitrarily in setting or changing conditions of employment.

Of course, the rights of employees established in contracts are limited by considerations of moral permissibility. Employers cannot give employees the right to violate the rights of others, for example, allowing employees to assign workers to life-threatening working conditions without their consent. Thus, in addition to employee rights explicitly agreed upon and established through contracts, employees have other rights. Many of these are related to the moral and civic rights that belong to individuals because of membership in particular societies. For this reason, to apply in working environments, these rights need not be explicitly articulated in employment contracts.

130. Again, for a fuller discussion of human rights in general, see Ropp and Sikkink (1999) and Nickel (1987).
131. For a cross-cultural analysis of justice, see Leung and Morris (2002).
132. For more on the nature of rights in general and their relations to engineering specifically, see Martin and Schinzinger (2010), Chapter 3.

Topics at the intersection of moral rights and employment might include fair compensation and hiring practices, equal opportunities for promotion, humane working conditions, protection against sexual harassment, degrees of privacy in working environments, and the nature of safe and healthy working environments, depending on the nature of the job. Topics at the intersection of civic rights and employment might include expression of religious preferences, protection against arbitrary dismissal, and participation in managerial processes. As with moral rights in general, considerable debate exists concerning how such rights should be applied to employment contexts. A great deal of discussion surrounds rights to privacy and nondiscrimination, for instance.

11.4 THE RIGHT TO PRIVACY

With the development of technologies allowing for employee monitoring, the right to privacy has become an especially contentious issue. When employees sign employment contracts, they clearly consent to limitations on their actions—just as membership in society limits free actions. A concern surrounds the extent to which employers are entitled to determine whether employees are adhering to these limits. For example, employees have duties to perform their jobs at acceptable levels, but the following questions can be asked regarding the limits of employer involvement: are employers entitled to monitor all employee actions? Can they monitor the behaviors of employees outside of work environments, for example, by testing for off-duty drug use? Should employers be allowed to limit the amount of time employees spend visiting the restroom, during working hours? Can they administer polygraph exams, lie-detector tests? Should employers be allowed to search the personal possessions of employees?

Although answers to many of these questions have been established regarding the nonworking parts of people's lives—through constitutional protections coupled with court rulings—they are not yet fully settled in employment contexts and call for ethical reflection. Writing protections for workers into employment contracts would be one way to resolve such issues. Even in dealing with employee collectives such as unions, however, employers have been disinclined to volunteer such protections. In addition to privacy, discrimination in the workplace is another hotly contested topic, lying at the intersection of employment, rights, and ethics:

- Do you think employers should be allowed to monitor the actions of their employee, regardless of job performance? Why or why not?

11.5 SPECIFIC CONSIDERATIONS OF DISCRIMINATION: RACIAL AND SEXUAL

The United States, for example, has long debated its history of racial discrimination—although such reflections are by no means either unique or confined to the United States, occurring in other parts of the world with regard to not only racial but also ethnic groups. Generally, it has been accepted that such discrimination in employment contexts is morally wrong. Practices of discrimination reject the fundamental equality of all human beings, instead making a characteristic irrelevant to work-related duties, race, a central factor in decisions regarding employment. This has led to charges of "reverse discrimination," discrimination against the majority population.

To correct for past injustices and raise minority populations to levels where they can compete equally, practices of "affirmative action" have been instituted. In these schemes, for example, if two workers are equally qualified for a position, then the one who is a minority would be hired. Insofar as the nonminority candidate might actually be more qualified—although both must be qualified—but denied the position in favor of the minority candidate, this has been considered reverse discrimination: in the attempt to right a past wrong, another wrong is committed, so the argument goes. More recently, a growing awareness exists that discrimination based on sex, religion, and national origins falls into the same category as discrimination based on race.

Again, if workers are to be treated equally, then this should be true with regard to characteristics irrelevant from the perspective of work-related duties. As touched on in the exercise at the beginning of this chapter, discussions regarding sexual harassment point toward this recognition. In sexual harassment a particular group is singled out for adverse treatment, resulting in negative effects on the abilities of members of that group to perform their jobs. Sexual harassment takes two main forms: first, actions are directed against a particular individual or group, for example, sexual advances; second, a sexually harassing culture/environment is established, for example, discussions or jokes of a sexual nature. Confusion regarding sexual harassment can arise from the nature of harassment.

Actions that demean others by turning them into objects—rather than treating them as beings deserving of respect—are obviously immoral. However, if sexual advances are welcomed or appreciated, then harassment has not occurred. Therefore, sexual harassment depends on subjective intentions and interpretations: the same action considered harassment in one set of circumstances might not be considered harassment in another. Women and men could have different interpretations of sexual actions in the workplace. Attempts have been made to establish objective rules that would limit behaviors. Given the subjective dimensions of behaviors, however, these rules will likely be inadequate. For this reason, issues surrounding harassment should be considered from not only a legal but also an ethical perspective:

- Aside from race, would policies like affirmative action be appropriate for other categories of personal and social identification? Explain your answer, listing and describing any other such categories.
- How should one react to seeing a coworker sexually harassed in the workplace? If you think your workplace is an environment of sexual harassment, what is the best course of action to take?

11.6 EMPLOYEE RIGHTS, IN GENERAL

In listing the rights of engineers, previously established duties can be used as a guide; Chapter 6 established the duties of companies in relation to stakeholders. Since duties entail rights and vice versa, when applied to employees as stakeholders, employee rights follow from the duties of companies.

Employees have the right to

1. be protected from unnecessary harm in their employment;
2. fair and just treatment by employers;
3. not be subjected to discrimination or harassment in their employment;
4. be treated by their employers based on merit;
5. have their contracts honored.

These rights are directly derived from corporate duties and, therefore, do not require further justification.

In addition to corporate duties, Chapter 6 listed employee duties. Turning back to this list reveals an additional employee right to

6. disobey illegitimate employment directives.

The other duties of employees establish the rights of employers rather than employees. Duties associated with these rights would, thus, receive further clarification in a text on stockholder—rather than employee—ethics.

11.7 THE RIGHTS OF ENGINEERS, SPECIFICALLY

Previous chapters have emphasized that, given their professional status, engineers are more than simply employees: they belong to a special category of employee with duties specific to their roles as engineers. Engineers have role responsibilities they should carry out in an autonomous fashion. To fulfill these special duties, engineers should also have specific rights, over and above those of other employees. Even if these rights are not explicitly formulated in employment contracts, in hiring engineers, employers implicitly agree to them. Just as general employee rights can be derived from corporate duties, so too can specific rights of engineers be derived from their duties to ensure trust from the public.

First, based on previous discussions of safety and whistle-blowing, as a special category of employees, engineers have the right to

7. blow the whistle (externally or internally) if, in their professional judgments, the physical safety of the public will be endangered as a result of their failures to act.

It should be remembered that, in accordance with previous discussions of whistle-blowing, the right to whistle-blow does not necessarily entail an ethical responsibility to do so.

Second, based on the duty of engineers to perform their work in an adequate fashion—so as not to endanger the safety of themselves or others—engineers have the right to

8. obtain the resources necessary to perform their assigned tasks competently.

Third, based on the relationship of engineers with the public, engineers have the right to

9. inform the public of engineering decisions that have the potential to seriously harm the physical welfare of the public.

In contemplating if, when, and how to exercise this right, it is important to keep in mind the complexity involved in balancing duties related to loyalty and confidentiality with those related to public safety.

11.8 ANOTHER RIGHT: INTELLECTUAL PROPERTY

One right not included in the above list—since neither does it directly affect the abilities of engineers to engineer more ethically nor does it directly follow from the duties of engineers—but that should be mentioned, is the right of engineers to property. Worldwide, schemes of intellectual property are becoming evermore complex. For this reason, on a global basis, the property rights of engineers should be outlined explicitly.

To encourage the development of new technologies, engineers should have the right to obtain just rewards from their achievements, including long-term gains resulting from their contributions to intellectual property. As "property" becomes an increasingly fluid notion—especially in international contexts—it is important that ownership rights be articulated. A final right of engineers would be to

10. fair compensation for their work, including the right to share equitably in gains resulting from their contributions to intellectual property.

Here the right to gains associated with intellectual property is based on the nature of technological development more generally, rather than engineering specifically. Articulating this right might be one of the more controversial claims of this text.

11.9 THE ENFORCEMENT OF RIGHTS

Before concluding this discussion of rights, it is necessary to briefly consider their enforcement. Unlike the duties of engineers, the preservation of rights requires an enforcement mechanism—since individuals and institutions are expected to respect rights. In life, rights are often violated. As discussed above, establishing legal mechanisms is one way to ensure the recognition and enforcement of rights. Drafting and enforcing laws can establish all types of rights—human, moral, civic, contractual, and so on.

Unfortunately, however, all too often those in positions of authority are the ones who violate rights. Therefore, additional actions are necessary. These often include boycotts, publicizing rights violations in news media, resigning positions, and so on. In addition to legal actions, these may or may not be effective. Just an awareness of and the ability to articulate the above rights, however, could further the ability of engineers to have their rights respected:

- Since moral and human rights are only recognized and enforced in some parts of the world, what determines the importance of rights worldwide? Do individuals and groups that recognize and enforce such rights have responsibilities to help other individuals and groups to attain this recognition and enforcement? Why or why not?

11.10   SUMMARY

As professionals, employees, and members of moral and civic communities, in addition to duties, engineers have rights. The nature of these rights in general, and specific rights, can be understood in terms of duties. Rights are the flip side of duties, and they can be conceived in terms of levels: the lower the level, the more basic and universal the associated rights, becoming more specific at higher levels. Insofar as engineers have obligations associated with the roles they occupy, engineers also have rights that should be ensured, allowing them to fulfill these obligations. In the sphere of employment, chief among such rights are those protecting individuals from discrimination—where characteristics of individuals and groups unrelated to job performance are used to make decisions regarding employment. Although a right to intellectual property is not directly related to public safety, given the present state of technological developments and social roles of engineering, engineers should enjoy the gains resulting from their intellectual property. Finally, although the enforcement of rights cannot be guaranteed, an awareness of rights by engineers could help to ensure they are.

REVIEW QUESTIONS

1. Explain the difference between positive and negative rights and the relationship of rights to duties/obligations?
2. In which way are rights related to autonomy? Under what conditions would it be ethical to restrict certain rights?
3. What are moral rights? Why do they "trump" other types of rights? Does one have to be human to possess moral rights?
4. In terms of what three levels can the rights of engineers be considered? Explain the relationship between these levels and the duties engineers have to public safety.
5. Why should the intellectual property rights of engineers be explicitly outlined?

6. List and explain the six employee rights discussed above and the ways they are derived from corporate duties.

7. Explain one way of insuring the recognition and encouragement of rights?

8. Define sexual harassment. Explain the two main forms of sexual harassment discussed above.

REFERENCES

Flores, A. (1980). Engineer's professional rights. *Issues in Engineering Journal of Professional Activities*, *106*(4), 389–396.

India Bans Captive Dolphin Shows as 'Morally Unacceptable.' (2013), Environment News Service. May 20. http://ens-newswire.com/2013/05/20/india-bans-captive-dolphin-shows-as-morally-unacceptable/.

Leung, K., & Morris, M. W. (2002). Justice through the lens of culture and ethnicity. *Handbook of justice research in law* (pp. 343–378). New York: Springer.

MacIntyre, A. (2013). *After virtue*. London: A & C Black.

Martin, M., & Schinzinger, R. (2010). *Introduction to engineering ethics* (2nd ed.). New York: McGraw Hill.

Nickel, J. (1987). *Making sense of human rights: Philosophical reflections on the universal declaration of human rights*. Berkeley: University of California Press.

Nussbaum, M. (2003). Capabilities as fundamental entitlements: Sen and social justice. *Feminist Economics*, *9*(2-3), 33–59.

Regan, T. (2004). *The case for animal rights*. Berkeley: University of California Press.

Ropp, S., & Sikkink, K. (1999). *The power of human rights: International norms and domestic change*. Cambridge: Cambridge University Press. Vol. 66.

Whitelaw, R. (1975). The professional ethics of the American engineer: A bill of rights. *Professional Engineer*, 37–41. August.

Appendix I

Global Engineering Ethics Principles Reviewed

I.1 BASIC ETHICAL PRINCIPLES FOR GLOBAL ENGINEERING

Based on their expertise, engineers should endeavor to

1. keep members of the public safe from serious negative consequences resulting from their development and implementation of technology;
2. ensure that fundamental human rights are not negatively impacted as a result of their work with technology;
3. avoid damage to the environment and living beings that would result in serious negative consequences, including long-term ones, to human life;
4. engage only in engineering activities they are competent to carry out;
5. base their engineering decisions on scientific principles and mathematical analyses and seek to avoid the influence of extraneous factors;
6. keep the public informed of their decisions, which have the potential to seriously affect the public, and to be truthful and complete in their disclosures;
7. understand and respect the nonmoral cultural values of those they encounter in fulfilling their engineering duties;
8. endeavor to refuse to participate in engineering activities that are claimed to reflect cultural practices but that violate basic ethical principles for global engineering.

I.2 ORGANIZATIONAL ETHICAL PRINCIPLES

Corporations should endeavor to

1. avoid producing unnecessary harms to those in and outside of their organizations;
2. ensure that all stakeholders of their organizations are treated fairly and justly;
3. ensure that all relevant laws and regulations are followed within their organizations;
4. protect members of their organizations against internal discrimination and harassment;

5. make all hiring, compensation, promotion, and termination decisions based on merit;
6. ensure that all legitimate corporate contracts are upheld.

I.3   ETHICAL PRINCIPLES FOR EMPLOYEES

Corporate employees should endeavor to

1. obey all legitimate, job-related directives;
2. perform their contracted duties on at least an industry-standard level;
3. uphold the principle of confidentiality in relation to knowledge gained in present and past employment;
4. avoid actions that harm the corporation in acting on behalf of the organization;
5. be honest in their business relationships with others;
6. enforce all organizational and employee ethical principles, when in positions of authority.

I.4   PRINCIPLES OF INVOLVEMENT FOR ENGINEERS

1. Principle of public participation—Engineers should seriously consider participating in public policy discussions regarding future applications of technology.
2. Principle of public education—Engineers should seriously consider helping the public to understand the applications of technologies in broader social, global contexts.
3. Principle of engineering engagement—Engineers should seriously consider becoming involved in helping to improve the technological futures of those less fortunate than themselves, on a voluntary basis.

I.5   RIGHTS OF ENGINEERS AS EMPLOYEES

Like all other employees, engineers have the right to

1. be protected from unnecessary harm in their employment;
2. fair and just treatment by employers;
3. not be subjected to discrimination or harassment in their employment;
4. be treated by their employers based on merit;
5. have their contracts honored;
6. disobey illegitimate employment directives.

As a special category of employees, engineers have the right to

7. blow the whistle (externally or internally) if, in their professional judgments, the physical safety of the public will be endangered as a result of their failures to act;

8. obtain the resources necessary to perform their assigned tasks competently;
9. inform the public of engineering decisions that have the potential to seriously harm the physical welfare of the public;
10. fair compensation for their work, including the right to share equitably in gains resulting from their contributions to intellectual property.

Appendix II

Steps in the Case-Study Procedure

1. **Identifying Ethical Issues**
   Ethics concerns actions that have the potential to seriously impact the lives of others, either directly or indirectly. List at least three ethical issues, posed in the form of questions.
2. **Narrowing the Focus**
   Completeness in analysis is preferable to superficiality. Choose one of the above issues on which to focus, giving a short justification for the importance of the issue chosen.
3. **Determining Relevant Facts**
   Facts related to case studies on engineering can be grouped into the following three categories: (1) material facts, (2) facts regarding individuals, and (3) facts regarding organization(s).
4. **Making Reasonable Assumptions**
   Facts will always be missing, although one can make reasonable assumptions. List any relevant missing facts and reasonable assumptions one can make regarding these facts.
5. **Undertaking Definitional Clarification**
   Go back through the previous steps, providing clarification regarding the use of terms and concepts that might be unclear, in relation to both issues and facts. Pay special attention to those that have "value connotations."
6. **Conducting Ethical Analysis**
   First, referring to principles for engineering, organizations, employees, and public involvement, list principles relevant to resolving the issue under consideration. Next, if conflicts exist between these principles, then decide which principles should take precedence. Is this hierarchy always applicable or only in this case? Finally, list any additional principles that would be relevant to resolving this issue.
7. **Reviewing the Process**
   As case study analysis is an iterative process, go back through the previous steps, seeing if there are other issues or facts that have been overlooked, terms/concepts that can be clarified, or principles that apply.

8. Resolving the Issue

Based on the previous steps, resolve the issue under consideration, answering the question you identified in step 2. Additionally, give a brief justification for your answer.

9. Identifying Practical Constraints

Although the answer given previously is ideal, do practical constraints exist that could reasonably be said to excuse either individuals or organizations from the answer given? If so, then list what these excusing conditions would be.

10. Avoiding Ethical Problems

Finally, based on the previous steps of the case study analysis, how might ethical problems have been avoided in the first place?

Appendix III

Guided Analysis: The Case of "Curious George"

To clarify the nature of the case study procedure, the following is a guided analysis of a hypothetical case of "Curious George."

III.1 THE HYPOTHETICAL CASE OF CURIOUS GEORGE

George Simon is a mid-level production engineer employed by Ajax Corporation. Due to a less-than-satisfactory relationship with his wife of 10 years, George tends to stay in the office until late in the evening, often just playing computer games. As a result of his late hours, he has earned a reputation as a hard working and productive employee who is regularly permitted to take extra long lunch hours. George uses these lunch hours to visit his mistress, to whom he gives office supplies from work, which she uses in a business she runs from her home.

One Dec. evening at approximately 11 p.m., George is staring mindlessly out of his office window when he sees an unmarked semi-trailer truck pull up to the loading dock. Curiously, he watches as the truck is loaded with industrial waste, which he recognizes as low-level radioactive waste products from the production line. Although this waste is handled by the environmental division of the company, George knows its disposal is usually contracted out to an environmental disposal company that operates according to a strict set of government regulations. He has also heard rumors that some in company management consider this to be an unnecessary waste of company resources, since proper disposal is a very expensive process. Recent changes in regulations have made the disposal of the waste twice as expensive as it was previously. This added expense occurs at a time when shareholders in the company have been pushing for a greater return on their investments.

Having nothing better to do—and not wanting to go home to his wife—out of curiosity, George decides to follow the truck. After a half-hour drive, the truck stops at a lake and dumps its load near the shoreline. The truck then drives off, and George, somewhat reflectively, drives home. There his wife is waiting up to criticize him, since there is not enough money to pay the mortgage. She wants to know where all the money goes every month, since there never seems to be enough to pay the bills. George slaps her across the mouth and decides to sleep on the couch.

The next day, playing around on his computer, George looks up the lake where he witnessed the industrial waste being dumped. To his surprise, he discovers that lake is the principal source of municipal water for the community in which he lives. George decides to encourage his wife to start a new fad diet, which involves drinking 2 gallons of water every day, and looks forward to a few beers over lunch with his mistress:

1. Identifying Ethical Issues
A number of ethical issues arise in this case, only some of which are directly related to engineering. Nonetheless, listing many of the issues helps to better think through the case. One can list issues in the order in which they occur in the case, remembering to formulate them as questions:

- Should George use company resources for personal entertainment?
- Should company personnel decisions regarding employees be made based on the number of hours spent at work?
- Should some employees be given extra privileges based on the quality of their work?
- Should George take company office supplies to his mistress?
- Should George's mistress accept Ajax supplies from George?
- Should George have a mistress?
- Should George base his actions on rumors?
- Should Ajax be more concerned about the cost of waste disposal or its environmental impact?
- Should Ajax dispose of radioactive waste through abnormal channels?
- Should George interfere with business matters outside of his own division in the company?
- Should George follow the unmarked truck?
- Should George take any action based on the information available to him?
- Should Ajax be held responsible for what a contractor does with its waste products?
- Should George slap his wife?
- Should George encourage his wife to take actions that might poison her?
- Should George drink alcohol at lunch on a workday?

The above list is reasonably complete, but you might discover additional ethical issues further along in the analysis, or you might disagree that all of the questions listed above are ethical in nature. Reasonable disagreement regarding how to interpret the materials dealt with here is to be expected and a natural part of ethical discourse.

Readers should note that the above list includes a mixture of ethical issues raised, including ones related to personal ethics, business ethics, and engineering ethics. Additionally, note that ethical issues can be expressed from a number of different perspectives, including those of George, the company, George's mistress, the environmental division, and the trucking company, among others.

2. Narrowing the Focus

Some of the issues above are so intuitively clear—such as George slapping his wife—that they might not warrant further discussion. In this regard, however, one should be careful not to move on too quickly. Not everyone necessarily shares the same point of view, where answers seem straightforward. That is why detailed consideration of even relatively straightforward questions can often help to clarify and justify decisions. Discussing ethical issues can often bring to the fore ambiguities and/or special circumstances concerning a situation, which have the potential to refine understandings of a case.

In deciding on a main ethical issue, there is also likely to be disagreement. Some might place the issue of harming others above all others, whereas others might focus on the actions of Ajax, while still others might pay attention to George as a person. Since the central concern here is with engineering, it makes sense to focus on an issue related to George's role as an engineer, since he is the only person the case identifies as an engineer. Interestingly, none of the questions raised above have this specific focus directly, although it is clearly an emphasis built into the case. Thus, it makes sense to reformulate one of the questions raised in step one to reflect this overall focus:

In his capacity as an engineer working for Ajax, is George ethically required to take action with regard to the knowledge he has gained concerning waste disposal, and, if so, then what action?

Notice that this question both clarifies a question posed above and places the issue in the context of engineering ethics. The issue is also formulated in a manner that will eventually necessitate the discussion of several of the other issues raised above. This formulation allows for the possibility of an integrated discussion—rather than simply a disparate consideration of a variety of different questions on a random basis, as often occurs in oral discussions of ethics.

3. Determining Relevant Facts

For the sake of this illustrative discussion, it is not necessary to list all of the relevant facts but, rather, initially to focus on only the major ones. The following list is divided into facts provided and facts not provided, but that would be important to know about the case.

Facts provided
- George is a mid-level production engineer.
- George works for Ajax Corporation.
- George has a good reputation at Ajax.
- George has a mistress.
- George showers his mistress with Ajax office supplies.
- George observes an unmarked truck late at night.
- George identifies the cargo as low-level radioactive waste generated by Ajax.
- George is not in the Ajax division responsible for radioactive waste.
- There is a firm that regularly handles Ajax's radioactive waste.

- There are governmental regulations for radioactive waste disposal.
- Waste disposal costs for Ajax have recently doubled.
- Environmental waste disposal is a great expense to Ajax.
- Ajax shareholders are requesting better returns on capital investment.
- The unmarked truck dumps the environmental waste into a lake late at night.
- George hits his wife.

Facts not provided but relevant
- Is Ajax engaged in an illegal process?
- Does the dumping have potential human health effects?
- What is the character of George?

4. Making Reasonable Assumptions

The next step consists in seeing if reasonable assumptions can be made about missing facts considered important. Each should be considered in turn.

First, if Ajax is not doing anything illegal, then the ethical question is not necessarily resolved, since dumping waste in the lake could still have consequences. However, if the dumping is illegal, then this might be a reason for George to take specific actions. The fact that dumping occurs late at night, with an unmarked truck, makes it reasonable to assume the action might be illegal, especially since the waste was dumped in a lake rather than an environmental landfill. Even if one assumes the process is illegal, however, it is still not clear that Ajax, as a corporation, is involved. Perhaps individual employees have initiated the dumping for personal financial gain. Hence, it cannot be assumed Ajax is engaged in illegality.

Second, we do not know if the waste disposal has serious health consequences. However, we can assume it would, since George wants his wife, whom he dislikes—also a legitimate assumption—to drink a lot of the water that comes from the lake. Additionally, since strict government regulations exist concerning the disposal of this waste, it would be legitimate to assume the government also judges that dumping has potential human health effects. Although the specific nature of the particular load of waste dumped is unclear, since the material falls under government regulations, this fact further legitimates assumptions regarding the negative health consequences associated with dumping.

Third, the need to answer the last question is unclear. The text previously emphasized the importance of action rather than character. However, character might be relevant to the determination of excusing conditions. Furthermore, the personal actions of George are given considerable emphasis, such that a focus on character could be expected. It is relatively clear from the facts given that George does not have an admirable character. This can be added to the list of assumptions, even if—later in the analysis—this assumption proves superfluous.

5. Undertaking Definitional Clarification

Value-laden terminology—such as "good," "bad," "just," or "unjust"—typically requires clarification. For example, assessing the character of George as bad might require clarification. In this case, however, "bad" character is sufficiently defined in terms of the actions of George. Another term that might influence the nature of the analysis is "mid-level," in reference to George's status as an engineer. If the word is understood as referring to his status as a "middle manager," then the role of the company would be involved in a way different from if the term was used merely to refer to George's function as an "experienced engineer." The latter seems more reasonable here, since the case does not refer to any supervisory role George occupies. This clarification would simplify the analysis.

6. Conducting Ethical Analysis

Here the first step consists in simply listing applicable principles. Since George is being considered in terms of his role as an engineer, the responsibilities related to that role are of primary relevance. Referring to Chapter 4/the list in Appendix I, these include principles related to public safety (1), human rights (2), environmental protection (3), and truthful disclosure (6). A closer examination of these principles reveals, however, that their application might be subject to some debate.

The question could arise as to whether the above consequences result from the implementation of technology by George. From the point of view of his job—engaged in the production process—the waste is, indeed, a result of his actions as an engineer, despite the fact that another division of the corporation is responsible for waste disposal. Hence, principles (1) and (2) seem to apply. It thus follows that principle (3) applies as well. Finally, if any further actions George might take in this situation are related to engineering, then principle (6) applies as well. From the point of view of engineering principles, George has duties to the public in relation to the waste disposal, since the facts and legitimate assumptions point to potential physical harm to the public. However, George is also an employee of Ajax, and his duties to the corporation should be considered as well.

Referring to the work from step 5, it should be noted that principle (6) of employee ethics has been eliminated from consideration, since the status of George is simply that of engineer rather than an authority. However, principle (3), on confidentiality, and principle (4), on harm, do seem to apply. On confidentiality, if Ajax disposes of waste unethically, then the company would not want this information publicized. Even if not, its waste disposal practices would probably lead to very bad publicity for Ajax. On harm, the situation is more complicated. At issue are two points: not only whether the company would be harmed in the long run if George took any further actions but also whether he would be acting on behalf of the corporation if George took further actions.

As is often the case in ethical analyses, here different considerations compete. Based on the brief discussion thus far, the engineering principles noted above clearly apply, while the employee principles mentioned are more problematic. Preliminarily, the judgment George should act on his knowledge—fulfilling his ethical responsibilities as an engineer—is justified.

7. Reviewing the Process

In considering the analysis thus far, perhaps it becomes evident that the status of George as he undertakes his late-night activities has been insufficiently considered. Specifically, although George was at work the night he witnessed the dumping, he was not actually working. Would his mere physical presence at his place of work be sufficient to evaluate his obligations in terms of employee status? Or, in following the truck, is he acting as either an employee or engineer? George seems to be motivated by mere curiosity rather than any responsibilities that follow from these roles. If his actions are considered in terms of these role responsibilities—and engineering considered a kind of social experimentation—then George would have a responsibility to follow up on the products resulting from his engineering/production activities, even waste products. Whether or not George is motivated by engineering ethical impulses, his possible actions fall within the purview of engineering actions and, thus, can be considered in terms of his role responsibilities as an engineer—even if not necessarily a general employee.

8. Resolving the Issue

Having clarified the remaining ambiguity above, the issue can now be resolved. As an engineer, George is obliged to take further action. While he has a duty of confidentiality in relation to his employer, that duty would not include confidentiality regarding illegal activities. To fully understand this claim, it is important to recognize that if the company is actively involved in the disposal of waste, then it is violating several organizational ethical principles, including principles (1), (2), (3), and possibly (6); again, refer to Chapter 6/the list in Appendix I. The right of the company against employees revealing confidential information is overridden by its unethical behavior. If the company were unaware of the illegal activity, then organizational ethical principle (3) would imply that it would want George to act.

However, that George is ethically required to act only resolves part of the ethical issue posed above. What action or actions should George take? Since George has an obligation as an engineer to protect the public and Ajax has an obligation not to harm the public, any action or actions that would protect the public are ethically justified. Thus, George could talk to his corporate superiors or go directly to the public, making a statement to the media, for example. Both of these actions could cause trouble for George, however, causing him to be perceived as a disloyal employee.

9. Identifying Practical Constraints

Realistically, George cannot be expected to take either of these actions, given the assessment of his character. Does the bad character of George serve as a condition to excuse his potential failure to act? The answer to this question is no, since a "bad" character is not the same as a "weak" character. One could legitimately assume a bad character is reformable, although a weak one might not be. If bad character served as a mitigating condition, then it would be nearly impossible to criticize the actions of bad people. Other types of practical constraints that could be considered include physical threats against one's life, the nature of corporate retaliation, and dire personal circumstances. None of these seem to apply in this case. Hence, the ethical requirement for George stands, even if he cannot be reasonably expected to act.

10. Avoiding Ethical Problems

George might have saved himself trouble—if the actions of Ajax were discovered and it was revealed that George knew about them—by not following the truck. However, this would have been at the expense of fulfilling an ethical duty. This avoidance maneuver would not, therefore, be a way of avoiding ethical problems. A better way of avoiding ethical problems might consist in staying informed about the process of disposing waste from the production process, working to help the company reduce waste disposal and/or production costs through greater efficiency, or ensuring one is working for a company and with colleagues who are ethical.

Index

Note: Page numbers followed by *f* indicate figures and *np* indicate footnotes.

Printed in the United States
By Bookmasters